T0310411

LEADING BIOTECHNOLOGY ALLIANCES

LEADING
BIOTECHNOLOGY
ALLIANCES
Right from the Start

ALICE M. SAPIENZA
and
DIANA STORK
with JOSEPH G. LOMBARDINO

 **WILEY-LISS**

A JOHN WILEY & SONS, INC., PUBLICATION
New York • Chichester • Weinheim • Brisbane • Singapore • Toronto

For ordering and customer service, call 1-800-CALL-WILEY.

Library of Congress Cataloging-in-Publication Data:

Library of Congress Cataloging-in-Publication Data is available. 0-471-18248-6

Printed in the United States of America.

10 9 8 7 6 5 4 3 2 1

CONTENTS

A SMALL INTRODUCTION
TO THIS BOOK

While we were finishing our writing, we struggled with the question of where to put a few introductory remarks that would help set the stage for the reader. We considered having a Preface, a Foreword, a very small first chapter. . . . Our final decision was: A SMALL INTRODUCTION TO THIS BOOK.

What the book is not. This book is not a comprehensive or exhaustive overview of alliance issues and problems, nor is it a compendium of advice to leaders of biotechnology alliances.

What the book is. This book is a tightly focused discussion about a set of issues that, we believe, are unique and critical to biotechnology alliances. It is a concise presentation of our ideas and insights (grounded in experience, history, and theory), intended to provoke people to think differently about alliances. This book is an elaboration of a process and our purpose:

Think differently. . . .
        Lead differently. . . .
                Make alliances work.

We present some evidence of the alliance "problem" and move on to examine the two sectors (pharmaceutical and biotechnology) and "typical" companies within them, to explain why there is such asymmetry between partners in most biotechnology alliances. It is this asymmetry, coupled with power differences between firms, that leads us to argue that the biotechnology side *must assume the greater leadership responsibility* in these alliances.

Against the backdrop of sector history and alliance context, our material is arranged chronologically, according to the timing of key events during an alliance. Our objective is that alliances be led right, right from the start. We begin with the issue of preparing to collaborate, and then we examine leadership issues and leadership roles over the alliance life cycle: reaching the first milestone, the middle stages of growth and maturity, and completion (or termination).

Our audience. We are writing for those who want to understand this organizational form of collaboration, who will participate in them, and who must manage and lead alliances well. First and foremost, the book is intended for leaders on the biotechnology side, because we contend it is their responsibility to make alliances work.

Our thanks. First, we thank the many biotechnology and pharmaceutical company executives, managers, and scientists who shared their alliance experiences with us, giving generously of their time. We want especially to thank the people of "Lucida," the pseudonym for a real company (our Chapter 1 case), whose members shared their alliance experiences with us, giving generously of their time over time. Second, we want to thank the anonymous reviewers for their extensive written feedback on an earlier draft of this book. Finally, we want to thank the editorial staff of Wiley for their patience, as we responded to feedback, conducted more interviews, wrote and rewrote sections and chapters, and, in the end, crafted what we believe is a better book.

If this book contributes to the success of just a few biotechnology alliances, we will consider it as having accomplished our goals.

LEADING
BIOTECHNOLOGY
ALLIANCES

PART I

TROUBLE IN ALLIANCE LAND

In the beginning stage of our research, we believed that we knew the issues and problems that needed to be addressed. We realized that biotechnology alliances had problems (as all organizations do) and that biotechnology alliance leaders faced challenges and hurdles (as all leaders do).

What we found was different from what we expected. The nature, magnitude, and frequency of problems surprised us. From our experiences in the field, as consultants and as researchers, we came to appreciate the extraordinary complexity of these organizational forms. Despite their proliferation, biotechnology alliances (most of them between a small and a large company) are very difficult to lead and manage well. Over and over, we heard people talk about the challenges in making alliances work. We witnessed the difficulties first hand, including the "untimely end" of a collaboration and the resulting demise of a company.

The first two chapters in this book present a backdrop of alliance problems at two levels of description and analysis. In Chapter 1 we provide a case study of the real (but disguised), particularly troubled alliance between *Lucida Biotech* and *Pharma Sciences*. As you read the case, you may be struck by how familiar the story is in a few or many regards, whether from your own first-hand experience or from the second-hand experience of others. In Chapter 2 we go beyond the Lucida–Pharma case to look at biotechnology alliances in general. We believe that most readers

1

will not be surprised by the magnitude of the general problem. Nonetheless, it is worth underscoring that the typical biotechnology alliance is troubled. Some end abruptly and prematurely; others struggle over time, expending unnecessary resources. Many face problems large enough to hamper their effectiveness, or even threaten their success.

CHAPTER 1

A CASE IN POINT:
THE LUCIDA–PHARMA ALLIANCE

Research alliances between small biotechnology companies and large pharmaceutical firms play a crucial role in helping to "close the innovation gap" in the biopharmaceutical industry; and, they are expected to be a major source of new therapeutics in years to come.[1] But, these efforts are very difficult to lead well. A typical research alliance or discovery alliance (we will use the terms interchangeably):

- Is fraught with uncertainty, ambiguity, and risk (typical of early stage research efforts)
- Involves dissimilar organizations (in terms of culture, size, power, and expertise)
- Is extremely important to the small partner (a failed alliance can be fatal to a biotechnology company)
- Is only modestly important to the large partner (these companies manage a portfolio of alliances with a number of biotechnology firms).

The task of leadership is even more complicated than the points above suggest, because the responsibility for keeping an alliance on course should

[1]Van Brunt, Jennifer. Innovation Drives Alliances, *Signals* magazine, www.SignalsMag.com, 1999.

not be shared equally by the partners. In fact, we have concluded that responsibility *should* rest on the shoulders of leaders in the biotechnology company. For a number of reasons that will be discussed in this book, alliances should be led and managed by the "little guys," even though it is the "big guys" who are experiencing the innovation gap and are concerned about their pipelines.

We begin this book with a real case, written from the perspective of the biotechnology partner. We want to introduce the people and the issues involved in the alliance between *Lucida Biotech* (the small partner) and *Pharma Sciences* (the large partner) right away, for several reasons. First, in and of itself, the story should be instructive. The case should provoke you to think about what happened, what was done, what might have been done differently, what you would have done, and so on. Second, we refer to this case throughout the book, for purposes of illustration and emphasis. Third, although this is the story of only one alliance, its characters, plot, and ending are all too familiar in this industry. The experiences of the *real* (but disguised) people in these *real* (but disguised) companies are, unfortunately, not unusual.

THE CASE OF A TROUBLED ALLIANCE

Lucida History

The Startup. In the early 1980s, Lucida Biotech was one of many companies emerging in and around "Genetown" and its world-renown universities. Lucida was formed when Dick Rosenbloom, an expert in a particular type of protein, was funded by venture capitalists hoping to "catch the wave" of promising biotechnology discoveries (refer to the cast of characters).

For the next 6 or 7 years, Lucida was essentially Rosenbloom's company. There was a group of six senior scientists—some of whom came from other biotechnology startups and the rest from universities—each in charge of a research program. They all began as project leaders and "advanced" to managers at the same time. This group, plus Rosenbloom, constituted the senior management of the company.

In 1988, the venture capitalists on the company board urged Rosenbloom to take Lucida public, but the banks strongly recommended that he hire a business person first. A new president with business expertise was

"Big Guys"
(big pharmaceutical companies)

"Little Guys"
(small biotechnology companies)

Cast of Characters

Pharma Sciences ("Big Guys")
Philip (Phil) Dean, MBA
  President and CEO since 1985
H. Ross Johnson
  Chief Operating Officer (hired in 1996)

Lucida Biotech ("Little Guys")
Geoff Pitchly, JD
  President and CEO (hired mid-1990)
Richard (Dick) Rosenbloom, PhD
  Founder (1982) and Chief Scientific
  Officer (CSO)
Mark Santoro, PhD, DSc
  Vice President, Research (hired in 1998
  from academia)

Stig Johanssen, PhD
  Vice President, Development (hired
  1992)
Janet Herman, MBA
  Vice President, Strategy (hired 1992)
William (Will) O'Brien, PhD
  Senior Scientist (joined company shortly
  after founding)
First scientist (program manager, hired in
  1998 from a large pharmaceutical
  company)
Second scientist (one of six senior
  scientists, with Lucida since shortly
  after the founding)

hired; however, he left about a year later, after taking the company through a successful initial public offering (IPO). Dick Rosenbloom found himself president, again.

Around 1990, money was tight, and some of the staff had to be let go. At about the same time, Geoff Pitchly, a lawyer from an international subsidiary of a big pharmaceutical firm, was hired as president. Rosenbloom took on the title of Chief Scientific Officer, equivalent in rank (at least in the way the company worked on a daily basis) to Pitchly. Rosenbloom was then working on a project of interest to a large health care company, not Pharma Sciences, and an alliance was struck between the two organizations. That deal was a collaboration in name only. No scientists from the other company ever worked with or even met any of the Lucida scientists. As described by a Lucida project manager, money was provided by

> . . . a conservative, solid, methodical company that regarded Lucida as someone hired to perform a service for them. Their idea of motivation was to fire the bottom 10% of their own people every year. But, they experienced a 20% compound annual growth rate for nearly two decades!
>
> There was never a meeting between our scientists and theirs. This was strictly a senior management deal. Their top management did insist on attending all quarterly meetings of Lucida's research program, and they brought along their consultants as experts. Those meetings were incredibly formal—the reports were so thick and detailed that we hated the approach of the quarterly meeting. It was very difficult to keep the Lucida scientists motivated, because everything stopped for 2 weeks while they assembled the report.

Compared with other biotechnology alliances, this deal was anomalous in two important ways. First, there was no scientific collaboration; second, the duration was almost a decade. Not until 1998 was the crucial clinical trial completed, demonstrating the hoped-for efficacy of the original compound. People in the company expected that this would become Lucida's first product.

Evolving Company Structure. One of Pitchly's tasks as CEO was to build a leadership team that could begin to craft a strategy for Lucida. A scientist who had been there almost since the inception of the company described what had been Lucida's approach as an *organ of the month club* strategy. Not atypical of other startups, Lucida's early R&D direction was not so much set in advance as described in retrospect. The company had expertise that was applied to whatever currently appeared promising. However, Rosenbloom expected that Lucida's expertise would be broadly useful, and he encouraged researchers to patent aggressively.

Although many were speculative patents, they did put Lucida far ahead of other biotechnology firms. These patents also caught the eye of the chief executive officer of Pharma Sciences, Phil Dean. Dean had been hired a decade earlier from a multinational medical products firm and sat on the

Board of Directors of Lucida, so that he knew Pitchly. (In addition to his professional relationship with Dean, Pitchly had a social relationship with Pharma's chief operating officer, H. Ross Johnson, who had been at Pharma for about two years.) When Dean recognized how strong Lucida's patent position was, he initiated the process that resulted in a formal alliance between the two firms starting in 1997.

Pitchly had hired vice presidents of strategy and of clinical development when he first joined Lucida. In 1998, a few months after the Lucida–Pharma Sciences alliance was signed, he actively recruited someone to lead research, with the encouragement of Rosenbloom. (It should be noted that Rosenbloom's strength was more in his conception of research experiments than in the actual leading of research scientists.) Following intense interviews with both Pitchly and Rosenbloom, Mark Santoro was hired as vice president of research, at the same level as the VP of development. Santoro came from an academic laboratory, where he was professor of molecular genetics at the medical school. He had a track record of bringing in large National Institutes of Health (NIH) grants, had an excellent reputation for leadership in the teaching hospital where his laboratory was located, and was interested in moving to industry.

Lucida Scientists. When he first interviewed at Lucida, Santoro was told about several people in the company, and one of Pitchly's comments stuck in his memory. Pitchly remarked that their lead bench scientist, Will O'Brien, was a difficult person and someone he would have to "deal with." O'Brien had been hired by Rosenbloom from the NIH and was considered a Rosenbloom protégé. Santoro assumed that Rosenbloom could not, or would not, manage O'Brien, and Pitchly appeared very unwilling to deal with him. He felt that one reason for hiring a VP of research was that *someone* needed to take on this job. The other remark Pitchly made was that O'Brien had expected Santoro's job; in fact, he had formally applied for the position.

When Santoro actually met O'Brien, it was not an auspicious beginning for their relationship. A luncheon was set up with him and two other scientists who had been there from the company's beginning. When they walked in, O'Brien did not meet Santoro's eyes, nor did he speak a word. Santoro tried to engage him in conversation, but only when O'Brien talked about a new research program did he become animated. However, the other two scientists did not appear to share his enthusiasm for the project.

Senior scientists at Lucida.

In fact, Santoro said that he saw little camaraderie among the three senior scientists—it was as if they worked at three different companies. They
all had their own stories, and each story was somewhat different from the
others'. Santoro described his interaction with O'Brien at that meeting as
"lack of recognition, lack of engagement, and downright avoidance."

After several months in the position, it became clear to Santoro that
what O'Brien cared for was science and creativity, not personal interaction. Although he talked about team spirit, he was not a team player. He
certainly was competent and very quick to make innovative connections.
But, Santoro noticed that once O'Brien had made up his mind about
anything—people, results, procedures—he never changed his opinion.
O'Brien had, he believed, a "black and white" way of approaching people
and work.

Santoro soon realized that Rosenbloom's history of running the company solo for so many years had resulted in an organization that essentially
revolved around him. "Power" to the scientists meant access to Rosenbloom, over or around the vice president of research and the vice president

of development. Santoro also noted that Rosenbloom was very forgiving in his relations with O'Brien:

> They've worked together more than a decade, and they have a nice pattern of relating to each other. Rosenbloom tolerates O'Brien because, I think, he can't confront anyone. He sees that O'Brien is intolerant of others' opinions. He knows that happens, because he is very articulate about what O'Brien's problems are. He's not blindered about them; but, he tolerates them and lets them happen.

Pharma Sciences History

Like other major pharmaceutical companies, Pharma Sciences was *big*, especially in comparison with Lucida. The company had offices all over the world and counted employees in the tens of thousands. Revenues were in the billions (US$), and the firm was involved in at least 20 alliances with biotechnology companies at any given time.

Senior managers were seasoned executives in health care and experienced at drug development. Phil Dean, CEO, was described in the press as a "tough and effective manager, demanding and getting results from those who work for him." He had been with the company for many years and had taken over the top position when Pharma was barely profitable. Dean sold off non-pharmaceutical businesses, built a highly respected management team, reduced the number of research programs in which R&D scientists were involved, and instituted strict budgetary controls. Another article described Dean as a "tireless worker with a penchant for midnight staff meetings. He formed 'productivity committees' to find areas where costs could be cut."

As a result of Dean's efforts, the company's profit improved and investors returned. Pharma Sciences became known as one of the most fiscally conservative of the majors, and executives maintained that they would not veer from a conservative course. Pharma Sciences also had a strict, quantitative approach to portfolio management. Development programs that did not meet the required and preset criteria were outlicensed, temporarily "shelved," or stopped outright.

Dean was appointed to the Board of Directors of Lucida Biotech in the early 1990s. He became impressed with the firm's aggressive patent strategy, and he believed that Lucida's expertise fit well with one of the large

Pharma Sciences research program. After several preliminary conversations with Pitchly, Dean sent representatives from Pharma Sciences to Lucida, and a contract was drawn up in late 1997.

The Lucida–Pharma Alliance

We began to interview people at Lucida in the spring of 1998 and continued through the fall of 1999. The first person we spoke with was Mark Santoro. He had joined Lucida about a half-year earlier, and he was concerned about the collaboration:

> Stig [Johanssen], our VP of development, was in charge of the alliance, but he had everyone in R&D reporting to him. When I was asked to attend meetings between Lucida and Pharma Sciences about 5 months ago, my assessment was that the alliance was headed for disaster. Stig had too many responsibilities and could pay little attention to the collaboration. Moreover, O'Brien played a large role in the work. There was a new program manager, but no one—including Stig—supported him and no one on the team paid attention to him.
>
> Ostensibly, there was a Lucida core team being coordinated by the new program manager, but because he was not supported, people simply did not show up for team meetings. Oh, they would show up at the biweekly meetings with Pharma Sciences, but they spoke independently, without vetting things with their group. Usually, Lucida scientists ended up arguing with each other in front of the Pharma scientists. This was just awful.
>
> Eventually, and with some difficult negotiations involving Stig, I was put in charge of the Pharma alliance. After a lot of work on my part, we are beginning to have a real team and a much more focused program of work here. The alliance is not yet where I want to see it, but it seems to be better than it was. Still, O'Brien functions autonomously.

About 2 months later, we spent time with Lucida's president (Geoff Pitchly) and VP of strategy (Janet Herman). For Pitchly, no alliance was problem-free:

> Alliances are always difficult, because there is never parity between the partners. There's no parity between Lucida and Pharma Sciences. They have sales. We don't! We don't have sales or royalties; we're not an operating company. We want to become one, though.

I have to credit Dean for a great deal of perseverance at Pharma, for a bull-dog approach to improving the company. He sits on our board, and I have a social relationship with Pharma's COO [Johnson], so there is a level of knowledge and respect between the companies. Pharma was interested in a partnership with us on a different project very early in our research. We still have a way to go, however.

Remember that, despite the relationships, the agenda for each company is different. And, it is still business. You can be friendly with people, you can respect them, but their business is to build Pharma and mine is to build Lucida. I think you can do that collegially; but, we had issues to be resolved at my level and other levels in the company.

I also worry that Pharma's priorities might change. What happens if another of their external initiatives takes off? We're just one of many collaborative research projects to them, some number in the large queue of resources.

Janet Herman was more optimistic:

Alliances work well if the two folks at the top are personally dedicated to it. For example, our relationship with Pharma works because Pitchly and Dean are dedicated. This is Dean's deal. When the opportunity was presented, there was skepticism on his management team, some discussion of why Pharma should do this, and so on. But, it was Dean's personal project, so they made an effort to see if the science would work. The fact that you have commitment at the top changes the interactions at the lower levels.

We remained in touch with Santoro, and about 6 months after this discussion we returned to Lucida. We spent time with two senior scientists, each of whom had major roles in the Pharma Sciences alliance. At this point, they viewed the collaborative interactions as troublesome and were attempting to understand what had gone wrong, and what was going wrong, as we spoke:

First Scientist (the new program manager, who joined Lucida at about the same time as Santoro)

When I look back to the initial meetings between the two groups of scientists, I realize that Lucida went in and did a 'data dump' and *then* Pharma scientists began to work out just what was involved. A number of the studies were not consistently reproducible. So, Pharma asked for some repeat studies in outside labs.

When those results came in, they were ambiguous. It took weeks before we could understand what was going on. Turns out there are longtime scientists in Lucida who are fonts of information on the compound. But, they sit in a meeting with Pharma people and bring something up and we say to ourselves: "Oops! I never heard that before!"

One of my counterparts at Pharma told me that he overheard Dean say: "If there were a lemon law for biotechnology, this alliance would be eligible."

The head of the project at Pharma became so bitter over the way these meetings went that he would not return our project manager's phone calls. Publicly, they called each other names and had a pissing contest in one of the meetings.

I spent a lot of time trying to rebuild relationships. The Lucida side of the team had weekly meetings to hash things through. Then, every protocol was reviewed by the whole project team, and both Lucida and Pharma people had to agree. After that, we started to get good data.

The biggest problem was that, instead of pulling back and reviewing our data thoroughly when they began, Pharma scientists just went full steam ahead. Then, all hell broke loose. It was: "full steam ahead;" then, stop dead. . . .

Second Scientist (one of the six senior scientists who constituted the early "management team" of Lucida)

Now that we're having problems, I can think of a number of earlier issues that concerned me. For instance, I think there was a certain amount of resentment at Pharma that management [Dean] went out and brought a molecule in. Johnson, their COO, was interested in our compound because of his prior experience in a related area. Also, he and Pitchly are friends, so Johnson knew about the project. But, that was not scientific or clinical experience.

Before we signed the deal with Pharma, several of our scientists went to a big symposium at NIH on [the relevant disease]. Interestingly, no one from Pharma went. We came back from NIH, and half of the scientists said: "This is daylight madness!" The other half said: "We'd better be very careful if we get into this work." I don't think the Pharma scientists realized what was involved, because they had not been to that NIH meeting.

Anyway, Pharma entered a partnership with Lucida, hoping to go from our bench results and our early in vivo work into the clinic quickly. We start to see some results that are inconsistent, but that is eventually straightened out.

I would also say that we did not manage the relationship terribly well. At the time, the work was being run by several people, but Pitchly said we had to have one person in charge of the alliance. That was a problem, because O'Brien thought he should be in charge.

O'Brien is very smart scientifically, but he is not managerially inclined! He believes that, if you're the boss, you tell people what to do and they do it. He's one of my best friends, and we've worked together for years. But, he has said to me: "I'm the only intelligent scientist in this company. Everybody else is a blithering idiot!" Now, I'm a scientist, and I'm his friend, but he says this with genuine sincerity.

So, O'Brien believed he should be in charge of the program, but Pitchly— and Rosenbloom—did not want to put him in charge. Improbable as this seems, Pitchly took charge of it himself.

We engaged in stealth management of the alliance for a while. The day before a meeting between Lucida and Pharma, I would put together the agenda and tell Pitchly what points people should address. Of course, during the joint team meetings, O'Brien was publicly dysfunctional.

We certainly did not come across as a coherent organization. Pharma wanted certain data. O'Brien had the data, but because he was not in charge, he was not going to give the data to the Lucida project manager.

I also believe we have problems, now, because this deal was done at the highest levels. I think, if you talked with Pharma scientists, they would say the deal with Lucida was forced on them by senior management. They had questions that were not adequately answered, and they did not have time to ask questions that should have been asked.

Because the deal was done from the top, Pharma management expected to go into the clinic immediately. When scientists from both sides got together, it became more and more evident that this program was not yet at that stage. I don't know if Pharma will have the staying power that we need.

A few months later, it became apparent that Pharma did not have the "staying power," and Lucida executives faced the painful termination of this alliance.

An Untimely End

Public statements made by the respective executives about the termination of the Lucida–Pharma Sciences deal reflected how important (or not) the alliance was to each partner. Although each company's annual report carried an announcement, the location of the announcement was an interesting commentary, in and of itself.

For example, a reader would have had to comb the Notes to Pharma Sciences' Financial Statements to find that a research alliance with Lucida had been formed in late 1997 and that $20M had been paid out in licensing fees. The one, brief paragraph concluded that Pharma Sciences had returned the "responsibility for [eventual product X] development to Lucida" in fall 1999, but retained the option of reassuming development for another calendar year. (In fact, they did not exercise that option.)

In contrast to the location of the Pharma Sciences announcement, the subject was addressed upfront by Pitchly, Lucida's president, in the introductory letter to shareholders. His letter stated: "The Pharma Sciences' agreement has been modified and, subsequently, promising new data have been discovered by Lucida researchers relevant to [Product X]."

In addition to suggesting differences in the importance of the alliance to the respective partners (reflecting the different financial dependence of each company on the alliance), the preceding statements also reflect the different impact of "bad news." At the time, Pharma Sciences was engaged in about 20 research collaborations with small biotechnology firms, reported product revenues in the billions of dollars, and had achieved a 50% increase in profits from 1997 to 1998. Lucida, on the other hand, had just completed one alliance with a large health care company; had no product revenues; and had *decreased their losses* by a few million dollars between 1997 and 1998. Obviously, Lucida was much more vulnerable to bad news about the partnership and the more dependent partner.

A short article in the press reported the following:

> The chief executive of Pharma Sciences said they decided to shift their R&D allocation to compounds with greater likelihood of near-term success. Lucida researchers had described the early in vivo work as promising, but both groups subsequently found ambiguous data from Phase I studies. Pharma Sciences was concerned that the ambiguity could delay even the design of Phase II trials by 12 to 18 months.

Another was more forthcoming, stating that Pharma Sciences had "abruptly withdrawn their scientists from the project . . ."

> As a result, Lucida's share price dropped precipitously by more than one-third. The value of the original contract for Lucida was as much as $100 mil-

lion over 3 years. Now, Lucida is no longer eligible for most of the $80 million in milestone payments.

The reason for termination did not appear to be the lack of promising leads, nor the failure to show efficacy in Phase I trials (the data were ambiguous, as opposed to negative), nor a change in Pharma Sciences' priorities (their annual report described ongoing research efforts in the disease for which Product X would have been a therapy). Rather, the ambiguity of data resulting in a possible delay of 12 to 18 months was unacceptable to the large firm's management.

Because of the immediate and precipitous drop in Lucida's stock price, the public market was not a source of additional funds. Although the executives tried, they could not find other sources of "creative financing." That, coupled with loss of milestone payments from Pharma Sciences, meant that certain research programs at Lucida had to be stopped. People were let go—in fact, the company was immediately cut in half.

The situation became grim. When the first draft of our book was written, Lucida remained in business; but, by the time we went to press, the company no longer existed.

The untimely end of Lucida.

CLOSING THOUGHTS

The experiences of scientists and executives in Lucida (and Pharma Sciences) are not unique. Over the 12 or so months that we were writing this case study, we were also interviewing numerous individuals in biotechnology and large pharmaceutical firms and hearing similar stories. Our book is based on the experience of our interviewees. It is also based on our own experience, as consultants to both large pharmaceutical and small biotechnology companies (Sapienza, Stork, Lombardino) and as a scientist in a major pharmaceutical company engaged in alliances (Lombardino). We have used conceptual material and a historical perspective to provide context for our insights, although we have made every effort to discuss concepts and theory in an accessible manner. Be assured, however, that the statements we make and the positions and perspectives we describe are indeed grounded in both experience and theory.

Our collective first-hand experience, conceptual background material, the wisdom and ideas of others, and so on, have led us to one major conclusion that we want to state upfront: Management and leadership of these alliances *should* rest squarely on the shoulders of people on the biotechnology side. Alliances should be led and managed by the "little guys," even though it is the "big guys" who are experiencing the "innovation gap" and who need biotechnology expertise beyond their own in-house R&D.

For many of you, the Lucida case will have raised new questions and new ideas about alliance issues and alliance leadership. We expect the same to happen in a number of the chapters that follow. Of course, we also expect to provide answers to at least some of your questions. But, the first step is to have new questions and new ideas.

Think differently . . .
　　　　Lead differently . . .
　　　　　　　Make alliances work.

CHAPTER 2

THE GENERAL CASE: MANY ALLIANCES, MANY PROBLEMS

Ernst & Young's 1994 annual report on biotechnology noted that alliances between small biotechnology and large pharmaceutical companies were so numerous that these "multiple relationships . . . [were signaling a] paradigm shift toward virtual integration" of the industry.[1] Many were research alliances, signed at the early stages (i.e., discovery), and a smaller number were at the stage of clinical development. Since then, the number of research/discovery collaborations has increased to the point that they now account for the majority of biopharmaceutical alliances:

> [The] predominant number of deals are signed at the discovery stage. . . . By 1997–1998, they accounted for almost two-thirds of all alliances. Clearly, big Pharma is betting on Biotech's drug discovery technologies as one way to close the innovation gap.[2]

Research alliances proliferated during the decade of the 1990s. Because not all alliances must be reported to (for example) the Securities and

[1]Ernst & Young. *The Seventh Industry Annual Report*, Ernst & Young LLP, 1451 California Avenue, Palo Alto, CA, 1994.

[2]Van Brunt, J. Innovation Drives Alliances, *Signals* magazine, www.SignalsMag.com, 1999.

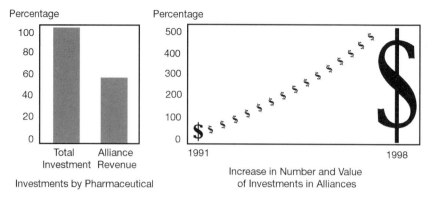

Importance of research alliances

Exchange Commission, any publicly available count underreports the number of deals. However, consider the following:

- Alliance revenue now provides more than half of the total capital invested in biotechnology.
- The most active pharmaceutical players in the alliance arena are engaged in 30 to 40 alliances each, at any one time.
- The number and value of investments in alliances increased by more than 500% between 1991 and 1998, and even greater reliance on collaboration into the twenty-first century is being predicted.[3]

The cyclical attractiveness of biotechnology initial public offerings (IPOs) and venture capital funding downturns continue to fuel an interest in partnerships between biotechnology startups and larger biotechnology and pharmaceutical companies. The increasing importance of alliances is also revealed by the shifting of the large partners' resources: *from* wholly in-house R&D *to* a sizable proportion—around one-quarter—on external collaborations. Even the largest biotechnology companies are shifting R&D resources to external collaborations. Amgen, which is itself partnered with pharmaceutical companies in marketing Amgen products, reported that about one-third of R&D resources were spent on alliances, some with biotechnology startups and some with universities. Clearly, research al-

---

[3]Coopers & Lybrand. *High-Performing Strategic Alliances.* Thought Leadership Series report, 1998.

liances with small, close-to-the-science companies are the source of many of the innovative ideas of today and the future. From societal and economic perspectives, it is important to make these alliances work.

MAKING ALLIANCES WORK

Despite their proliferation, alliances are problematic. The tensions and difficulties revealed in the Lucida–Pharma case study, in our interviews, and in our observations are consistent with findings published by Coopers & Lybrand and by *Signals*, the online magazine associated with the Internet-based Recombinant Capital site. According to a Coopers & Lybrand report, "between 30% and 40% of [biotechnology] alliances fail to meet the partners' expectations. . . . [Executives] are concerned about the ability of their organizations to successfully manage alliances."[4] Using Recombinant Capital data on biotechnology alliances, *Signals* reported that, consistently, about 10% of biotechnology strategic alliances "come to an untimely end" each year, because:

- "A research program has just not produced any promising product leads;
- a product has failed to perform in clinical trials or at the FDA;
- or the larger partner has reshuffled its research priorities, to the exclusion of the collaboration."[5]

Using these data, we could conclude that between and one-third and one-half of all biotechnology alliances are not working well. But, in fact, these data probably under-estimate the extent of the problem. Imagine how many interpersonal or collaborative difficulties (as opposed to scientific difficulties) an alliance could be facing that would not be recognized by senior alliance executives. If one could get accurate figures about those problems, we expect the result would be that well over half of all biotechnology alliances are in trouble.

Of course, there will always be scientific and technical setbacks in R&D, and invention and development of state-of-the-art products will fail sometimes. In fact, the closer to the "cutting edge" these projects are, the

[4]Coopers & Lybrand, 1998.
[5]SignalsMag.com, 5/30/99.

Many alliances are not working well.

higher their rate of attrition. But, successful collaboration depends not only on the solution of scientific and technical problems but also on the leadership skills of the people involved. One Eli Lilly executive (quoted in the Coopers & Lybrand report) said: "The biggest challenge is making alliances work."

How can scientists and executives ensure that, even in the face of scientific or technical setbacks (such as the ambiguous data Lucida presented to Pharma Sciences), the partners will value the relationship and its future possibilities sufficiently to work through the problems of the present?

This long question has a long answer, which is the focus/purpose of this book. We believe that you must think differently about alliances in order to lead them differently and to make them work better. We believe that to think differently, alliance leaders need new lenses through which to view alliances and new frames for understanding them.

We begin by reviewing the importance of biotechnology, and we present several propositions for why biotechnology alliances are, by their very nature, more problematic and difficult to lead than alliances in most other industries. Then, we set out our argument for concluding that making alliances work is the responsibility of leaders on the biotechnology side.

THE IMPORTANCE OF BIOTECHNOLOGY

In all industrialized nations, the high-technology sector has grown rapidly and now accounts for a sizable component of total national output. The success (or failure) of high-technology firms can influence the economic advantage of a nation. That is why biotechnology, like microelectronics, has been deemed a *strategic* technology in the economic sense of the word. The U.S. Department of Commerce listed biotechnology among 12 emerging technologies believed to offer singular economic potential.[6] The National Critical Technologies Panel reported to then-President Bush that biotechnology was among 22 technologies essential to the economic competitiveness of the United States.[7] More recently, the National Science Foundation included biotechnology as important "for economic competitiveness . . . and contribution to the quality of life."[8] Even the Organisation for Economic Cooperation and Development described biotechnology as "hot."[9]

Biotechnology has become a major source of new therapeutics and will continue to be in the foreseeable future. The pharmaceutical industry association noted:

> The 350 biotechnology medicines in [all phases of] development provide more compelling evidence that the biotechnology revolution . . . is not only the wave of the future—but very much the promise of the present.[10]

Because of its global economic potential, biotechnology has been widely promoted by means of industrial policies and other facilitating structures. In the United States, where the technology originated, a number of laws were passed in the 1980s to promote private sector investment. Public Law 97-34 created one of the primary means of financing startup biotechnology companies. This law provided tax relief for large, profitable firms (e.g., pharmaceutical companies) that invested in Research and Development Limited Partnerships (RDLPs) with biotechnology firms.

The tax relief incentives of RDLPs were reduced by the Omnibus Reconciliation Act, but biotechnology research collaborations continued to grow. Today, alliances have become the principal vehicle by which these

[6]Department of Commerce. *Technology Administration Report on Emerging Technologies*, Washington, DC, 1990.

[7]*National Critical Technologies Panel Report*, Springfield, VA: U.S. Department of Commerce, 1991.

[8]National Science Foundation. *Science & Engineering Indicators*, Washington, DC, 1999.

[9]OECD, Economic Outlook No. 66, Washington, DC, 1999.

[10]Pharmaceutical Research and Manufacturers Association. 350 Biotechnology Medicines in Development, www.phrma.org/charts/biointro.html, 1998.

(strategic) therapeutics are developed and diffused. A typical biotechnology company lacks the skills and resources to take an idea through the arduous processes of clinical development and beyond. The tremendous risk and cost of Phase III trials and then marketing a product can only be borne by the large companies. Furthermore, the accelerating pace of the cognate sciences requires that all possible sources of new ideas be tapped. Finally, large pharmaceutical (and the so-called first generation biotechnology) companies need a sizable pipeline of innovative products to sustain earnings growth. A portfolio of biotech alliances at the research stage plays an important role in meeting large companies' competitive needs.

SPECIAL LEADERSHIP CHALLENGES

As generic structures for organizing collaborative efforts, alliances are not new. However, a short-lived type of alliance has emerged in various high technology sectors, including biotechnology.[11] Such alliances are characterized by compressed life cycles. We have found that they also have sharper peaks of uncertainty (imperfect or incomplete information) and ambiguity (competing or conflicting perspectives).

More traditional alliances have a life cycle measured in years, sometimes in decades. Today's high-technology alliances have a life cycle that may be measured in months. An extreme (but not unique) case is the alliance between Parke-Davis (a pharmaceutical company) and Ribozyme (a biotechnology company). It lasted four months, from 12/97 through 3/98.[12] A compressed life cycle is characteristic of biotechnology research alliances and presents one special leadership challenge: There is not much time to "get it right." These projects often represent the quick hit, long ball philosophy of R&D, as opposed to the multiyear philosophy characterizing projects pursued in-house by big pharmaceutical companies.

Another special challenge arises from the nature of discovery itself. Discovery efforts are undertaken to produce something that is novel. The process is filled with uncertainty, ambiguity, and risk. Scientists involved at this early stage are often driven, ego involved, and not particularly adept

---

[11]Spekman, RF, Isabella, LA, and MacAvoy, TC. *Alliance Competence: Maximizing the Value of Your Partnership*, New York: John Wiley & Sons, 2000.

[12]Data are from Recombinant Capital data base (recap.com).

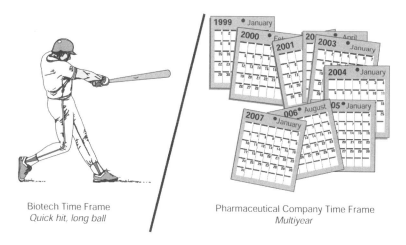

Biotech Time Frame
Quick hit, long ball

Pharmaceutical Company Time Frame
Multiyear

One leadership challenge—differences in R&D philosophies.

or interested in working as part of a team in the company. They have been described by one experienced leader as follows:

> Self-starters, neither seeking nor needing supervision or management . . .; risk takers with dislike for the status quo; not often obvious team members; [holding] strong convictions, forcefully expressed; [with an] ambitious focus on their scientific pursuit not their personal careers.[13]

Because company culture, structure, and systems influence the kind, level, and frequency of creative behavior, a leader of discovery efforts must be adept at providing organizational and interpersonal supports and at removing organizational and interpersonal impediments. Both the technical and interpersonal realities of a research alliance must be managed in a way that does not lessen the likelihood of productive creativity.

The organizational realities of biotechnology discovery alliances present yet a third special challenge. Two or more companies, each with its own customs and cultures, are engaged in *virtual collaboration*. No laboratory space is shared. Rather, experiments are performed in parallel as well as sequentially, but at different sites. *Actual collaboration* only occurs (or not) at team meetings, when people interact face-to-face or voice-to-voice.

---

[13]Fitzgerald, cited in Thorne, P. *Organizing Genius*, Oxford, UK: Blackwell Publishers, 1992.

LEADING WELL: BIOTECH'S RESPONSIBILITY

Given this context—the economic importance of biotechnology, and the special challenges of leading biotechnology research collaborations—it is easy to see why effective leadership is both crucial and difficult. Our intent, as we began this book, was to help scientists and executives from large and small companies work together productively and effectively. This is still our intent. Initially, however, we expected to recommend (as many authors writing about alliances have done) that both partners should share the responsibility for managing the alliance dynamics.

What we have learned, and now vigorously propose, is counter to the prevailing wisdom and has been expressed by one biotechnology executive as follows:

> The smaller companies invariably have less 'say' in the alliance. But, they have to do more to keep the alliance on track, and that is hard work. The alliance is a much bigger piece of their daily life, and it is much more important that it go well for them than for the partner. The alliance is *much more* important to the smaller company (vice president of research, biotechnology company).

The reality is that biotechnology strategic alliances are not relationships among equals. Because the alliance is more critical to the survival and success of the smaller partner, as Lucida's experience revealed, the relationship dynamics should be primarily the responsibility of leaders on the biotechnology side. Although we found alliances in which the biotechnology side was shouldering the burden of leadership, it was often without understanding why that should be so. Indeed, we often found some resentment among biotechnology leaders about why there was not a sharing of the leadership responsibilities.

Over the course of this project, we have come to appreciate why such imbalance in responsibility should be the case. Our conclusion that relationship responsibilities must fall more heavily on the shoulders of leaders in the biotechnology company arises from two facts. First, knowledge frontiers relevant to biotechnology are moving quickly, and the smaller firm is selected for its scientists' competence in the fast-moving knowledge base. Second, there is great disparity in size and power between the partners. The small organization is rather like a "gnat on a wildebeest" (in the words of one of our interviewees). Because of the critical role of the science, and the importance of alliance revenues to biotechnology firm survival, *we contend that leaders in the smaller firm should take on greater responsibility for the relationship.* We also acknowledge that taking greater responsibility is very difficult for the small company. With relatively few people to choose from, individuals may not be well-matched to the alliance leadership and management tasks they are meant to accomplish.

Explicitly assigning more responsibility to one partner, rather than to both equitably, contradicts contemporary alliance advice. Much of the current literature suggests—implicitly, in some cases—that both partners should (as in a marriage between relative equals) work to sustain and grow the relationship. But, as we listened to people from both sides, we were struck by themes of difference and imbalance. Repeatedly, people talked about the differences between the small, startup company culture and the large firm culture. The science was different, the people were different, the processes were different, the time horizons were different, the values were different, and so forth. We were also struck by many

The gnat on the wildebeest.

references to imbalance in alliance dynamics. We heard the word "bully," we heard about holding the purse strings, we heard the phrase "noose around our neck," and we heard "we feel like we're begging."

Because imbalance is a fact of life in these relationships, our leadership recommendations are directed mainly to leaders in the small company. It is up to them to make the alliance work. Of course, alliances are relationships, even if lopsided, so our insights and recommendations will be valuable to leaders in both small and large companies. As large firms continue to be involved in increasing numbers of alliances, those with reputations for being good collaborators will gain competitive advantage over those with the opposite reputation. Such companies will become viewed as "preferred partners," when small companies bring them new ideas.

ORGANIZATION OF THIS BOOK

The experiences of Lucida executives and scientists touch on many of the leadership issues that will be addressed in the remainder of this book. We do not begin by tackling them directly; rather, we step back to look at the evolution of the pharmaceutical and biotechnology sectors (Chapter 3). Their different histories produce contrasting sector cultures; they also help to explain the general dissimilarity between sectors and the general similarity among companies within each sector. Aspects of Lucida's culture (e.g., an opportunistic approach to strategy) have been influenced by the biotechnology sector culture. Similarly, aspects of Pharma Sciences' culture (e.g., fiscal conservatism, portfolio management) have been influenced by pharmaceutical sector culture.

In Chapter 4 we describe other substantive differences between typical alliance partners. Contrasting cultures plus differences in (for example) size between pharmaceutical and biotechnology firms produce asymmetric relationships and *very complicated alliance dynamics.* Alliance leaders must recognize that complicated dynamics are a "fact of life" in these collaborations.

Alliance leaders should understand sector history and the resulting general cultural and other differences between types of partners. In order to make any particular alliance work, they must also lay the groundwork to support collaboration (Chapters 5 and 6). Chapter 5 presents the steps that should be taken within the biotechnology firm. For example, it was obvious to the Lucida scientists, after the fact, that the research structure and systems were inadequate when the deal was signed:

> . . . At the time, the work was being run by several people. . . . Pitchly and Rosenbloom did not want to put O'Brien in charge. . . . Eventually the VP was hired, but he knew nothing about [Product x] biology. . . . We certainly did not come across as a coherent organization.

Chapter 6 describes the corresponding assessments that should be conducted with regard to the large partner (individual and organizational due diligence). For example, what is the intended strategic fit of the biotechnology invention? What is the reputation of the large firm as a collaborator? Which individuals in the partner company are likely to be working with the biotechnology scientists, and what are their educational backgrounds and work experience?

Although we cannot know for certain, the observations of Lucida Biotech scientists that their counterparts at Pharma Sciences may have felt "forced" into the deal suggest that addressing such issues may have been omitted when the alliance got under way. The task of identifying Pharma scientists likely to be working on the program should have led immediately to ensuring that those people were committed and involved from the start.

The prework necessary for effective collaboration includes a number of important activities, including careful planning for the first meetings of the combined team (Chapter 7). Without careful planning and attention to what may seem to be minor details, the first meeting can go badly. Consider what a Lucida scientist recalled:

> When I look back to the initial meetings between the two groups of scientists, I realize that Lucida came in and did a 'data dump' and *then* Pharma scientists began to work out just what was involved. A number of the studies were not consistently reproducible. So, Pharma asked for some repeat studies in outside labs.

Once this preparation is complete and the first formal meeting has successfully occurred, the next major hurdle facing leadership is reaching the first milestone (Chapter 8). The Lucida–Pharma alliance was terminated before that milestone was completely met (the press reported that Lucida was "no longer eligible for most of the $80 million in milestone payments"). We cannot give technical or scientific advice about working toward the milestone, but we can address the team issues that might have an impact on achieving the milestone. For example, the same Lucida scientist noted the following:

> The head of the project at Pharma became so bitter over the way these meetings went that he would not return our project manager's phone calls. Publicly, they called each other names and had a pissing contest in one of the meetings.

After the milestone is met, there are other issues that complicate the remaining stages of the alliance life cycle (Chapter 9). In each stage, there are differing levels of uncertainty and equivocality that have implications for team dynamics and are generally predictable. What can be predicted, of course, can be anticipated, planned for, and intelligently managed.

In Chapter 10 we discuss completion of the alliance, as well as termination, and describe a postalliance debriefing process to ensure that "lessons learned" are institutionalized. One of Lucida's executives stated that: "We do a deal so rarely that we don't do nearly as good a job as we might." We contend that even rare deals produce useful lessons, and leaders must codify the lessons and make them available to everyone as needed. Listening to each individual at Lucida, we found a wealth of experience that could become part of a shared knowledge base. Some of this knowledge could have been transmitted and shared between people directly and some of it could have been captured by an electronic expert system and made available on the company intranet. At the least, small companies should have a special debrief meeting, during which the senior management team reviews the just finished alliance, paying special attention to leadership issues and leadership roles. Lessons learned from a just completed (or terminated) alliance will make the next alliance work better.

Chapter 11 summarizes what we believe is needed for alliance effectiveness. We provide a roadmap for leaders to follow before, during, and after an alliance. Chapter 12 is a brief coda to the Lucida case study.

Think differently . . .

        Lead differently . . .

                Make alliances work.

PART II

ASYMMETRIC RELATIONSHIPS, LOPSIDED RESPONSIBILITY

Biotechnology alliances are relationships, but not between equals. They are relationships forged between very dissimilar organizations, in terms of culture, size, power, expertise, and so on. Alliance leaders need to understand that such differences are unavoidable and that they will complicate the alliance dynamics.

Biotechnology companies have their origins in the academic world; pharmaceutical companies have their origins in the chemical industry. Their beginnings are different, as are their paths to the present. In Chapter 3 we present a brief history of the two different sectors (biotechnology and pharmaceutical), to frame some of the cultural differences between alliance partners.

In addition to contrasting cultures, pharmaceutical and biotechnology companies have many other differences that have consequences for alliance dynamics and thus for alliance leadership. We explore these differences, some so large we call them disparities, in Chapter 4.

Taken together, Chapters 3 and 4 help to explain the asymmetric relationships we have found characterizing these collaborations. Such asymmetry is not easy to manage from the biotechnology side. The very survival of the biotechnology company often depends on the alliance outcome, while the other partner is the one with the money and the power.

Regardless, to make alliances work, people on the biotechnology side should bear the larger burden of leading and managing the relationships.

To a large extent, the two chapters in this section identify some of what we term "hardship conditions" facing leaders. These conditions apply to virtually all biotechnology alliances. Understanding this reality should lessen the chances of finger-pointing and the "us" versus "them" thinking that lays blame too easily on the other side.

> Contrasting cultures, along with differences and disparities between pharmaceutical and biotechnology companies, contribute to complicated alliance dynamics.

CHAPTER 3

CONTRASTING CULTURES

The paradigm shift toward virtual integration of the biopharmaceutical in-dustry might imply that the industry is integrating smoothly. From a height of 10,000 feet, perhaps, the terrain may look smooth. "On the ground," we know this is not the case; problems appear to be the norm rather than the exception. This is troubling, because an unsuccessful relationship between a small and a large company can result in catastrophic organizational fail-ure. For example:

- Initially, Lucida reduced its staff by 50%, because Pharma Sciences terminated the alliance; soon, the company ceased to exist as a sepa-rate entity.
- Shares of Synaptic (its real name) dropped nearly 50% when Eli Lilly terminated a joint research program.
- ImmuLogic (its real name) was dissolved after Hoechst dropped a for-mer Marion Merrill Dow collaboration of several years with the biotechnology firm.

We found that the alliances we studied, like the one between Lucida Biotech and Pharma Sciences, present formidable leadership challenges, many of which related to differences in culture (differences in norms, values, beliefs, etc). In some cases, those challenges grew into disastrous consequences, as in the above examples. In other cases, those challenges *only* complicated the dynamics, impeded progress, and generally made life difficult for alliance members and leaders on both sides.

When companies with very different cultures work together as alliance partners, we expect problems. And, indeed, alliances between biotechnology and pharmaceutical companies often show signs of cultural strain. Yet, many of the typical cultural problems we find in biotechnology–pharmaceutical alliances are both understandable and predictable, given the different sector histories. Biotechnology companies have their origins in the academic world, and pharmaceutical companies have their origins in industrial chemistry. Their beginnings are different, as are their paths to the present. We also discovered similarities within sectors; leaders in each type of company face many of the same problems and issues. The individual players and companies may change, but the alliance dynamics seem remarkably similar. Companies within a sector share many cultural characteristics and features, even though each has unique attributes. Across sectors, however, companies of each type are culturally very different.

We want to emphasize that the contrasting cultures of the biotechnology and pharmaceutical *sectors* contribute to complicated alliance dynamics and create special challenges for alliance leaders. Because the sectors have such an important influence, we are devoting this chapter to exploring sector history and resulting sector culture. We begin with a brief historical overview of the pharmaceutical and biotechnology sectors, and we explore some of the consequences of contrasting cultures for the leaders of strategic alliances. Understanding the history of the two sectors should help leaders better understand their own company culture and that of their partners. What biotechnology executives and scientists may have previously seen as problems with "them," they may now see as examples of cultural strain. Thinking differently entails (among others) reframing the alliance as a relationship across a cultural divide. Leading differently should follow.

BIOTECHNOLOGY AND PHARMACEUTICAL COMPANIES

Two different sectors . . .
Two different histories . . .
Two different cultures . . .

THE PROBLEM DEFINED

One of our interviewees characterized what could be called the "culture problem":

> There are two different cultures involved in these collaborations, each with different goals and expectations of the relationship. The effectiveness of each culture is based on premises that are actually antithetical to each other, but now each [side] wants and needs to work together (pharmaceutical company research director).

What he meant by "two different cultures" are the values, beliefs, and norms generally characterizing the pharmaceutical and biotechnology sectors. In the experience of everyone we interviewed, the *average* biotechnology firm was believed to be very different from the *average* pharmaceutical company (and vice versa, depending on the person's employer), although it must be emphasized that this average exists only as a construct.

We use the image of three concentric circles of culture that matter in the everyday dynamics of all biotechnology alliances. The outer circle is what we are calling sector culture. Pharmaceutical and biotechnology sectors have different histories. Participants in these sectors share experiences and learning. Over time, an identifiable and recognizable culture emerges at this level.

The next circle is organizational culture: how things are done in a particular company, the taken-for-granted routines, everyday decisions and actions people engage in without second thought, and so on. Organizations, like sectors, have different histories as well. Lucida was in many ways a biotechnology company like other biotechnology companies; it was also uniquely Dick Rosenbloom's company. Company norms reflected, in part, Rosenbloom's way of dealing with issues and people, particularly the chief scientist, Will O'Brien. As Mark Santoro said: "Rosenbloom tolerates O'Brien because, I think, he cannot confront anyone." As a result of this organizational culture, we would expect to find avoidance of conflict in Lucida. Similarly, Pharma Sciences' routines of portfolio management were described as "strict" and "quantitative," reflecting that organization's history (under Phil Dean) of strict budgetary control and fiscal conservatism.

The inner circle is discipline culture. Particular norms and values, ways of thinking, and ways of acting characterize every discipline (a discipline

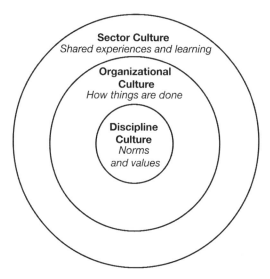

Three different levels of culture in an alliance.

is, after all, shared experience and learning). Molecular biologists approach problems in ways that are different from those of organic chemists. Neither approach is necessarily better or more correct; they are, simply, different because of shared experience and learning. The disciplines bring "different values, perspectives, and ways of seeing the world" that are vital to creative invention and the development of biopharmaeutical products.[1] When leaders can take advantage of this discipline diversity, promoting intellectual challenge and wide problem-search, the outcome will be more creative than if diversity is quashed.[2] On the other hand, if leaders are unable to harness the benefits of diversity, discipline differences can compromise the outcome.

Recognizing and understanding the organizational and discipline levels of cultural differences is certainly important for effective leadership, but those differences must be set in the context of sector history and culture. This history has a profound effect on the values, attitudes, and beliefs of scientists and managers in firms within that sector. The major

[1]Stork, D. Not All Differences Are Created Equal; Not All Should Be Managed the Same: The Diversity Challenge in Pharmaceutical R & D, *Drug Development Research*, 43: 174–181, 1998.

[2]Whatmore, J. *Releasing Creativity*, London, UK, Logan Page Ltd., 1999.

league pharmaceutical companies ("big pharma") have been operating since the nineteenth century; small biotechnology firms did not emerge until the 1970s, and most emerged in the 1980s. Big companies have a long track record of global research, development, manufacturing, sales, and marketing operations spread over a number of therapeutic areas. Small firms have a recent, predominantly local, and narrowly focused track record of discovery efforts. Only the three or four largest can undertake development and marketing of products, and only within one or perhaps a few therapeutic areas. Perhaps most importantly, "big pharma" evolved from the chemical industry, while biotechnology evolved from academia.

We begin with the evolution of chemical and dye firms to the modern pharmaceutical organization. Next, we describe the transformation of academic biomedicine in the United States and the corresponding transformations in academia and in industry that produced the biotechnology sector. We conclude with some of the consequences of history for alliance leaders.

EVOLUTION OF THE PHARMACEUTICAL SECTOR (THE INFLUENCE OF GERMAN SCIENCE)

The earliest producers of medicines were simple, individual suppliers.[3] From the fourteenth to about the mid-nineteenth centuries, they included the apothecary, the physician, and the herbalist. The first apothecary shop was opened in London in 1345. The Apothecary Guild was established in 1617 as an autonomous trade group, having been part of the Grocer's Guild since the twelfth century. Medicines ("drugs") were based on ancient formulae and consisted of complicated mixtures of herbs, some with hundreds of ingredients. Called "galenicals" from the original formulations of Marcus Aurelius' physician, Galen, these "drugs" were prescribed well into the nineteenth century—with no scientific evidence of efficacy, safety, or therapeutic intent for these agents.

During the nineteenth century, emergence of the formal disciplines of organic chemistry, human biology, and other quantitative sciences led to a more sophisticated approach to the use of medicines. Individual suppliers still prepared their own products (which included galenicals, botanicals,

---

[3]McLaughlin, T. *Note on the U.S. Pharmaceutical Industry*, Boston: Pew Curriculum Center, Harvard School of Public Health, 1986.

and organic chemicals like diethyl ether) and distributed them directly to the customer, resulting in a wide range of drugs and quality. It was during this period that, from a number of informal and ad hoc relationships, the modern pharmaceutical sector emerged.

The relationship between Paul Ehrlich and the German chemical firm, *Fabwerke Hoechst*, can be considered a prototype of alliances. It consisted of a relationship between a company and a single academic scientist. (This model emerged in the United States after World War I and continues today. What we observe most recently is not the abandonment of this form but the emergence of another, more common and increasingly important form.) Ehrlich, like other physicians and scientists of the time, was using dyes to stain tissues. He conducted experiments at Hoechst, using their methylene blue, and in the process refined his therapeutic strategy. Ehrlich's success prompted the strategic decision in German and Swiss chemical firms to include a pharmaceutical business:

> The development of chemotherapeutic agents thus became a logical opportunity for those chemical companies which were already engaged in the synthesis of dyestuffs. Within a few decades, all the major companies along the Rhine . . . started pharmaceutical operations.[4]

German university and chemical company scientists collaborated widely on both pharmaceutical research and university science curriculum, and that example found its way to America.

The international prominence of German pharmaceutical companies had no doubt been the result of several factors, but crucial among these were focus on the scientific method and emphasis on advancing the present state of scientific knowledge. Parke, Davis and Company was the first U.S. firm to set up a research institute; Lilly and Upjohn soon followed suit. (Formerly, these development and manufacturing companies had no in-house research capability.)

At the same time, American medical training was being reformed according to the German model, which emphasized rigorous technical preparation and the application of laboratory science to clinical practice. A similar reformation took place in the science programs of American universities, based on German scholarship. The Johns Hopkins graduate program, founded in 1876 and modeled on the German system, played a

[4]Drews, J. Drugs Made in Germany, *Int. Views of Politics and Science*, V1, 1988.

leading role in disseminating that system in the field of chemistry.[5] Unlike their German colleagues, however, Hopkins graduates had little interaction with industry. Some of the land-grant universities of the Midwest had ties, but they were the exception. One reason for the paucity of alliances was that only a few U.S. pharmaceutical companies had research facilities. Although American companies had begun to build a research foundation in the first two decades of the twentieth century, real growth was stimulated by World War I.

One consequence of postwar expansion was that alliances between American university and industry scientists began to flourish (as they had in Germany decades earlier), supported by funds coming from the enormous new industries:

> [The] creators and inheritors of the giant fortunes that were coming into being on the basis of such chemically-oriented industries as oil (Rockefeller), steel (Carnegie), explosives (DuPont) and photography (Eastman) were not slow to endorse scientific research on a wide and inclusive basis. . . . [In the field of chemistry there emerged] a rich network of graduate scholarships, faculty consultantships, research grants, equipment awards, and other incentives by which academic-industrial linkages were developed and sustained and research and innovation fostered.[6]

U.S. pharmaceutical company relationships with universities and medical schools came to resemble the earlier German example. Merck recruited an academic to head its pharmacology laboratory in 1932 and then established close connections with the University of Pennsylvania Medical School. A number of other companies (e.g., Upjohn, Pfizer) established close relationships with scientists at the University of Wisconsin School of Medicine.[7]

Events during World War II had a important influence on the pharmaceutical sector. American and British governments and American pharmaceutical companies collaborated to supply penicillin to the Allied troops. That effort had the unintended but beneficial effect of demonstrating that

---

[5]Geiger, R. *To Advance Knowledge*, Oxford, UK: Oxford University Press, 1986.

[6]Thackray, A. University-Industry Connections and Chemical Research: An Historical Perspective, in *University–Industry Research Relationships*, National Science Foundation, Washington, DC, 1982.

[7]Swann, J. P. *Academic Scientists and the Pharmaceutical Industry*, Baltimore: Johns Hopkins University Press, 1988.

the underlying biological mechanisms did not have to be understood. If it worked, it worked. . . . At the same time, trying to understand the action of an effective medicine could advance basic knowledge and might lead to more effective medicines.

Wartime investment in research personnel, plant, and equipment changed instrumentation and drug production techniques. To meet the military needs for the drug, bulk production was required. Because it was more economical for the manufacturer to supply drugs in the form in which they reached the ultimate consumer, brand names became important differentiators. Money spent on advertising and promotion of medicine rapidly increased toward the proportions that prevail today. After the war, a number of chemical companies merged with drug firms to market their products to civilians. For instance, Merck manufactured bulk drug-grade chemicals before the war but afterwards acquired Sharp and Dohme, one of the largest drug producers dating from the Civil War.

Another important influence on this sector was the progress made in organic chemistry in the first half of the twentieth century. That progress in understanding, coupled with growing knowledge of receptor theory (pointed to by Ehrlich, in fact), resulted in more rational drug design and synthesis within the pharmaceutical company. Drugs could be synthesized to act in a specific way (such as blocking a receptor), rather than being identified after the fact from random screening.

EVOLUTION OF ACADEMIC BIOMEDICINE (THE INFLUENCE OF NIH)

During the early decades of the 1900s, the U.S. government played a minuscule role in organized biomedical science. When the Hygienic Laboratory, precursor of the National Institutes of Health, was signed into law by Roosevelt in 1902, its budget was about 5% of the budget of the private Rockefeller Institute.

Real growth in public sector funding of American biomedical research began a few decades later. After years of debating the role of government, Congress passed the National Cancer Institute Act in 1937. After World War II, President Truman signed into law the National Heart Act, creating a second institute of health (hence, National Institutes of Health, or NIH). The stimulus for government involvement in biomedical science was the success of the atomic energy research at Los Alamos. The objective for the

NIH "was simple: If we can harness the energy of the atom, why can't we conquer the killing diseases . . . ?"[8]

Postwar expansion of NIH resulted in the dominance of government support of biomedical research in U.S. universities and medical schools and the growing renown of U.S. academic biomedical science. Research continued to be conducted within the NIH, but these institutes became the principal channel by which federal funds were distributed to universities, especially medical schools and teaching hospitals.

By the mid-1960s, annual federal biomedical research appropriations had reached $1 billion, with about three-quarters going to academic research facilities (*NIH Data Book*, various years). But, the postwar economic boom soon slowed because of the Vietnam conflict. In the mid-1970s, the OPEC crisis and recession further dampened biomedical funding growth rates. Between 1966 and 1976, NIH appropriations grew at only 7% (compounded annually), signaling to academic scientists that this empire was downsizing.

By 1979, more than $7 billion (current dollars) were being spent annually on biomedical R&D in the United States. Of this, 60% was spent on academic biomedicine and funded by the federal government, 40% through the NIH. Medical schools, of which there were 120 or so, received about $1 billion from the federal government. The larger teaching hospitals routinely had multimillion-dollar research budgets. Most of the remaining R&D monies consisted of industry funds. Pharmaceutical association member firms (then called the Pharmaceutical Manufacturers Association, with about 140 companies) spent $2 billion on in-house biomedical R&D.

A consequence, albeit unintended, of strong federal support of biomedical research was the near-disappearance of alliances between industry and academia. Individual academics continued to act as consultants to industry, but they did not depend on industry for support of their research. Pharmaceutical companies had supported about 10% of academic biomedical research in the early 1950s, but this dropped to little more than 1% in 1975. Overall, U.S. biomedical research activities, academic and industrial, had grown to be the largest in the world. University and industry alliances, however, had diminished. Government (NIH) funds supported research at medical schools and teaching hospitals like Massachusetts

[8]Strickland, S. P. *Politics, Science & Dread Disease*, Cambridge, MA: Harvard University Press, 1972.

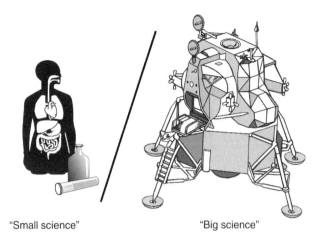

"Small science" "Big science"

General Hospital (MGH). The pharmaceutical industry funded and conducted its own research within company facilities.

After World War II to about the mid-1970s, organized biomedical science in the United States consisted essentially of the following:

- Scientists in academia working alone or in small teams, within the confines of their medical school, teaching hospital, and university department, on so-called basic research[9] (and serving as consultant to corporate R&D laboratories)
- Scientists in government working alone or in small teams, within the confines of their institutes, on so-called basic research
- Scientists in the pharmaceutical sector working alone or in small teams, within large corporate laboratories, on so-called applied research for the purpose of discovering drugs
- Government acting as primary source of funds for basic biomedical research in the public sector.

In comparison with efforts in the physical sciences, biological science was considered "small" science.[10] Biological technology was, similarly, "small"

[9]Peter Medawar defined basic research as that with "no substantive goal . . . [only a direction] toward . . . the 'truth.' " Medawar, P. B. *Advice to a Young Scientist*, New York: Basic Books, 1979.

[10]DeSolla Price, D. *Little Science, Big Science . . . And Beyond*, New York: Columbia University Press, 1986.

technology (e.g., microscopes and test tubes, again in comparison with, say, the huge instruments employed in Los Alamos atomic energy research). In addition, the academic and industrial sectors were believed to be clearly distinguishable in terms of (1) stage of research (i.e., basic was viewed as very different from applied), (2) objectives (public good or private gain), and (3) funding source (public monies or commercial returns). Relationships between academic scientists and the pharmaceutical industry were, as noted, rare. Those that did occur were very much an outgrowth of shared intellectual interest, often serendipitously encountered.

TRANSFORMATIONS

Biomedical Research

In Shakespeare's play, Ariel sings:

> . . . *Nothing of him that doth fade*
> *But doth suffer a sea-change*
> *Into something rich and strange . . .*
> *The Tempest*

Quite literally, biomedical research was transformed into something rich and strange by discoveries in molecular biology and "genetic engineering" (*biotechnology*). So revolutionary were the findings that James Wyngaarden, then director of the NIH, contended there had been a sea-change in biomedical research. The metaphor was apposite; biomedical science and technology, industry structure, and industry alliances were transformed.

Judah Folkman commented: "Biology used to be like botany. You could read a textbook 50 years old and not much had changed. Now, textbooks are outdated in 2 years."[11] Compared with the rate of progress of the past century, it was as if biomedicine's common and well-worn "bones" had turned to rare and precious "coral" in the blink of an eye (Ariel's song). Revolutionary, discontinuous discoveries triggered a rapidly advancing knowledge frontier in the biological sciences. Like a wave, insights gained about one phenomenon cascaded into other areas. Questions that may have been asked earlier but remained unanswered or only partially answered were now illuminated.

[11]Sapienza, A. *Preparing for Globalization: Takeda Chemical Industries' Central Research Division*, Boston: Pew Curriculum Center, Harvard School of Public Health, 1989.

The feedback loops between science and technology and between invention and innovation were dramatically shortened. Medicinal chemistry had been the technology of pharmaceutical discovery since Ehrlich collaborated with Fabwerke Hoechst. Almost 100 years later, this chronology and these events occurred: In 1973, Herbert Boyer of the University of California at San Francisco and Stanley Cohen of Stanford discovered how to introduce a piece of DNA from one organism to another (rDNA). In 1976, Boyer and Robert Swanson, a venture capitalist, created a company called *Genentech*. Thus, only 5 years after Boyer and Cohen's academic work, recombinant human insulin was created. Shortly thereafter, *Humulin* was marketed by Genentech.

These scientific accomplishments, although revolutionary, were less exciting to the public than the financial alchemy of Wall Street, which appeared to produce precious coral and fine pearls in minutes. As a reporter described, when Genentech held its initial public offering in 1980 (just 4 years after its founding), "thirty minutes after the market opened, the underwriters had sold their stock [at $35] and disbanded . . . Twenty minutes later the now freely trading stock hit eighty-nine dollars a share . . . a record for an initial public offering."[12]

Without exaggeration, those twenty minutes changed the biomedical industry and set off a rapid proliferation of a new type of alliance. These collaborations did not emerge ad hoc, nor were they brought about by shared intellectual interest. Rather, they were based explicitly on perceived competitive advantage. As the following discussions of transformations in academia and industry illustrate, the biotechnology sector emerged, and alliances became *strategic*.

Academia

We noted above that, from the 1950s through the 1970s, collaborations between U.S. academic and industry scientists were rare. By 1980, formal research relationships between American university scientists and multinational pharmaceutical companies were so numerous that the U.S. government requested an inventory and investigation.[13] Between 1979

---

[12]Teitleman, R. *Gene Dreams: Wall Street, Academia, and the Rise of Biotechnology*, New York: Basic Books, 1989.

[13]National Science Foundation. *University–Industry Research Relationships: Myths, Realities and Potentials*, Washington, DC, 1982.

and 1985, the compound annual growth rate of industry funding of the top 15 research universities in the United States was twice that of NIH funding.[14]

Some insight into the changes occurring in academia can be gleaned from the minutes of a 1975 meeting of the Scientific Advisory Committee of Massachusetts General Hospital (MGH) on research funding. The institution's research budget had grown considerably (nearly $20 million), but faculty emphasized that the slowdown in NIH funding rates made "greater diversification of the funding base critical." One of the committee members noted the following:

> . . . Although it seems clear that industry should not be looked to as a major new source of financial support for basic research activities, it appears likely that there will be areas of concurrent interest shared by industry and academic institutions.[15]

The committee agreed that the goals of industry and academia might not be so far apart. George Hitchings, then a Burroughs-Wellcome scientist who would later share the Nobel Prize in medicine, commented that profitable discoveries were the most creative ones and that basic research (the purview of university) was valuable to the industry. By the end of the meeting, the decision was made to encourage ventures between MGH and the pharmaceutical industry.

Certainly, academic and industrial scientists had collaborated before. Now, not only the numbers but also the drivers of alliances were different:

> From the university's point of view, the special appeal of the burgeoning industrial connection is quite simple—money. Federal support of basic research has been gradually declining for the past decade, and the situation has now been measurably worsened by the dismal state of the economy and the Reagan Administration's determination to reduce government spending.[16]

[14]Sapienza, A. *Technology Transfer: Pharmaceutical Industry Research Relationships with American Universities*, Boston: Harvard School of Public Health report to the Andrew Mellon Foundation, 1988.

[15]Sapienza, A. *Gaining a 'Window on the Science': The Hoechst–MGH Relationship*, Medford, MA: The Fletcher School of Law and Diplomacy, case study, 1990.

[16]Culliton, B. The Hoechst Department at Mass General, *Science*, 216, 11 June 1982.

From "publish or perish," the slogan could now be heard throughout academic biomedical research: "find money or perish." Said an academic researcher in 1989:

> We really have to pull in the research money. Publications are crucial, but if you publish without bringing money in, you will not be around very long. You have to bring money in.

There were still concerns about the impact of industry support on academic freedom, and another monetary issue became troubling as well:

> The *new* research collaborations differ in several respects [from earlier examples]. For one, the time lag between basic discovery and application or product development has been shortened. . . . But the primary difference is that these *new* collaborations are based on developing a product, and the expectation is that it will yield profits. Although there is nothing wrong with the profit motive, it does mean that a new motive has been entered into the biomedical research collaborative equation.[17]

Industry

From fewer than 20 biotechnology firms in 1979 there were more than 150 in 1983, about half of which were created in 1981 with Genentech as example. Almost every one of these companies began with one or a small number of academic biomedical scientists and venture capital funding. These scientists became general managers and part owners of their company, often with a very attractive financial package in addition to the stock.

Robert Nowinski was not atypical of the academic entrepreneurs who emerged as part of the "sea-change" that created the biotechnology sector. After receiving his doctorate from Cornell's Sloan-Kettering division, he moved to the Fred Hutchinson Cancer Institute. Working on monoclonal antibodies that had been discovered in the United Kingdom by two Cambridge scientists, Nowinski became convinced of their role in cancer research and therapy. With Genentech's example, Nowinski and two venture capitalists formed Genetic Systems in 1981. When Genetic Systems was

[17]Wyngaarden, J. *Address to Conference on Competition and Cooperation in the Biomedical Industry*, Boston: Harvard School of Public Health, 1989.

acquired by Bristol-Myers, Nowinski profited. And when he left Bristol, he started another biotechnology company with the same venture capitalists who had funded the first.

Walter Gilbert, a Nobel-Prize-winning molecular biologist and Harvard professor, was one of the founders of Biogen. Gilbert left Biogen in 1984 to go back to his academic position, while retaining membership on Biogen's Board of Supervisory Directors and Scientific Board. Later, he left his academic position and helped found another biotechnology venture.

Between 1979 and 1983, nearly $1 billion was raised in public offerings by the largest biotechnology companies formed by these academic entrepreneurs. Between 1979 and 1985, startup companies also raised nearly $0.5 billion in private equity, from venture capital as well as from numerous research and development limited partnerships (i.e., alliances) with pharmaceutical companies. By the end of the 1980s, nearly 1000 biotechnology firms constituted the biotechnology sector.

The "sea-change" in biomedical research, of course, affected pharmaceutical research as well. Medicinal chemistry had been the technology of discovery essentially since Ehrlich worked with Hoechst in the late 1800s. From a technology life cycle perspective, medicinal chemistry was in the mature stage of the logistic curve in the 1970s. This was reflected by a secular decline in numbers of new chemical entities produced by the industry. Continued increases in R&D investment (up to about 10% of revenues) had made the pharmaceutical industry one of the most research-intensive, high-technology sectors in the world. Although expenditures on R&D had nearly quadrupled after World War II, however, new chemical entities produced never reached their peak of 1967.

Biotechnology represented a new technology of discovery that was in the growth stage of a new (and metaphorically higher) logistic curve by the early 1980s. This new technology brought about changes in industry structure and the nature of competitive forces.[18]

The long stability and profitability of the pharmaceutical industry was based in part on very high barriers to entry, perhaps the highest being regulation. Drugs discovered by medicinal chemistry required an average of 10 to 15 years for development, completion of clinical trials, and regulatory review. Early biotechnology products faced a different scenario. Although biotechnology drugs constituted only 2% of drugs to be reviewed

---

[18]Sapienza, A. R&D Collaboration as a Global Competitive Tactic—Biotechnology and the Ethical Pharmaceutical Industry, *R&D Management*, 19(5), October 1989.

by the Food and Drug Administration (FDA) in 1986, they accounted for 20% of the approvals.

It must be emphasized that these first products received expedited reviews, because they were "low hanging fruits." They were compounds produced naturally in the body but now genetically engineered in the laboratory. Because they were *not* new entities, proof of safety was not a major issue. Production by biotechnology companies of natural proteins was relatively easy. The second, third, and fourth generations of biotechnology inventions are much more difficult to produce. Those "fruits" that do emerge are quickly selected, often termed "cherry picking," for further development within alliance efforts.

Another important barrier to entry, membership in the knowledge network, had been eroded.[19] Simply stated, the pharmaceutical companies were not connected to the *right* networks. Pharmaceutical leaders had paid little attention to the revolution in molecular biology during the 1960s and 1970s, because some senior managers believed that biotechnology was too speculative. Other pharmaceutical executives admitted that "many large companies, like ourselves, were . . . slow off the mark, way behind in the biotechnology race," because of a complacency reflected in the following interview from 1988:

> In the long run, one or two of the fledgling biotech companies might succeed, and I hope they do. But eventually some of the "big boys" will build the technology in-house and produce an appropriate number of novel, patentable, biotech products and ensure their profitable survival. It will be fascinating to compare and contrast the growth in business and profits of firms who invested heavily in biotechnology—like Genentech—during this "hype" phase of the technology.[20]

Historically, the commercial successes of the large pharmaceutical companies derived mainly from small molecules and synthetic drugs. Changing that philosophy came slowly. The established pharmaceutical companies thus became very dependent on biotechnology firms that were tightly connected to the new knowledge network, through the academic entrepreneur who founded the company. Soon, the industry was characterized by a web

---

[19]Burstall, M. L., and Senior, I. The Community's Pharmaceutical Industry, Luxembourg: Office for Official Publications of the European Community, 1985.

[20]Sapienza, A. *UK Biotechnology and the Research Links Between ICI and the University of Leicester*, Boston: Pew Curriculum Center, Harvard School of Public Health, 1989.

of alliances between established pharmaceutical and startup biotechnology companies. Genentech itself had between 15 and 20 alliances with large companies in the mid-1980s.

Although not quite on the scale of physics, biology in the 1980s was coming to resemble "big science" and "big technology" because of this integration. A U.S. National Science Foundation report predicted, in 1982, that the scale was not likely to decrease:

> First, product and process improvement and innovation in some industries have evolved to levels of complexity that demand understanding of fundamental physical and biological phenomena . . . and thus require much higher . . . use of basic science. . . .

> Further, incremental advances in narrowly focused technical areas, characteristic of much industrial development in the past, are giving way to the use of a broad range of science and engineering disciplines on complex, often ill-defined problems, or exploitation of new analytical capabilities. Hence, it is becoming increasingly difficult for any one industrial laboratory to fully encompass the requisite expertise.

> [Finally,] it is quite unlikely that any one company could hold and maintain a leading edge on technical advance in a given area [because of] momentous changes in [the biological sciences].[21]

The Sectors Today

Overall, high-technology (i.e., research-intensive) industries are driving global economic growth. The National Science Foundation reported that output from four such industries—computers, pharmaceuticals, communications equipment, and aircraft—grew from about 8% of global production in 1980 to 12% in 1995. Of the latter four, "only the U.S. pharmaceutical industry managed to retain its number one ranking throughout the 15-year period. It was also the only U.S. high-tech industry that had a larger share of the global market in 1995 than in 1980."[22]

Research-based pharmaceutical companies were expected to have invested $24 billion in R&D in 1999 and to have sales of about $127 billion.[23] Mergers such as Ciba-Geigy and Sandoz (to form Novartis), Hoechst,

---

[21]National Science Foundation. University–Industry Research Relationships, 1982.

[22]National Science Foundation. *Science & Engineering Indicators*, Washington, DC, 1999.

[23]Pharmaceutical Research and Manufacturers Association website, phrma.org, 1999.

Marion Merrill Dow, and Rhone-Poulenc Rorer (to form Aventis), and Pfizer-Warner-Lambert have made "big pharma" even bigger. The biotechnology sector has grown as well. In 1999, these companies were expected to have invested nearly $10 billion in R&D and to have revenues of nearly $19 billion. (Most of these revenues come not from sales but from other payments, such as alliance payments.) Market capitalization of this sector in the United States was estimated to be close to $100 billion. About 1200 companies constitute this sector in the United States, of which only a few hundred (300 or so) are public and account for most of the R&D and revenues. There are approximately the same number of firms in Europe. Unlike pharmaceutical operations, which are funded by sales of products, biotechnology operations are funded by venture capital, stock, and (primarily) alliances with big pharma:

> The number of $100 million deals increased to record levels with individual deals setting new highs. This 'marriage of convenience' is likely to continue, with the pharmaceutical industry gaining access to technologies to provide the next blockbuster drug, and biotech companies receiving, in some cases, life-sustaining cash.[24]

SOME CONSEQUENCES OF SECTOR HISTORY

> Biotechnology companies have different goals and behave consistently with these goals. A strategic alliance with the pharmaceutical industry is not easy; each sees the world from a different vantage (biotechnology scientist).

The vantage from which people in each sector "see the world" is influenced by the history of that sector. Companies in the pharmaceutical sector evolved from very-large-scale chemical manufacturing organizations. In addition to a long history of R&D and manufacturing competency, established pharmaceutical companies are skilled in promotion of their product—a skill dating from World War II. Biotechnology companies evolved from academic scientists funded by venture capital (and *venture* captures the spirit of the sector). They have a short history of scientific and technological competence:

> The origins of drug research . . . are in a "chemical worldview," and the cultural characteristics of chemistry at the time when the chemical, and soon

---

[24]Ernst & Young. *Biotechnology Industry Annual Report*, 1998.

Pharmaceutical and Biotechnology Worldviews

Pharmaceutical

- Big pharma
- Major league
- Shareholder value
- Rich product pipelines
- Legendary profit growth
- Blockbuster products
- Brand-name recognition

Biotechnology

- Entrepreneurial
- Research boutique
- Survival index
- Burn rate
- Creative financing
- Value-added innovation

thereafter the pharmaceutical, industry arose have left their mark upon the procedures and culture of this industry. This is an extremely rigorous culture of precision and objectivity, but also of hierarchical dependency, discipline, and subservience. . . .

. . . In contrast to chemistry, molecular biology stems from a primarily democratic, liberal, indeed libertarian social order, in which formal hierarchies play a much smaller role, while on the other hand, personal development and freedom are more important than in the society of a century ago.[25]

These worldviews, how people in each sector see the world (and how differently), are reflected in how they speak. We provide illustrations, first, from articles in the business press.

Sector culture is also revealed when people speak informally and spontaneously. We will illustrate with a few verbatim comments on *strategy*, *structure*, and *scientists.* These are not meant to stereotype companies or sectors, in any judgmental way, but, rather, to illustrate the differences between sectors more vividly.

First, consider the following excerpts about organizational strategy and biotechnology alliances, taken from our interviews with executives in pharmaceutical companies:

. . . We pursue a careful strategy of asset redeployment, and we target our research programs to selective therapeutic areas in which we have a strong franchise.

[25]Drews, J. *In Quest of Tomorrow's Medicines*, New York: Springer-Verlag, 1999.

. . . We ask: "Can we sell this? Will managed care buy this?"

. . . There are many business questions we ask about biotech: "Will an alliance be more profitable for us in the short term? Is there clear business logic to the alliance? Are the numbers OK?"

Contrast these with the following response by a senior manager in a biotechnology company:

Strategy? You start with dreams . . . that's what strategy is in this business. We're not selling products but our vision of what science can provide and what that might mean to investors.

Second, note that pharmaceutical executives describe company structure in terms reflecting large size as well as a number of standard operating procedures:

. . . We have therapeutic area teams, with representatives from strategic marketing, medical, and so on. Before something goes into development, they have to make recommendations to the portfolio committee.

. . . We have an office in the United States and one in Europe, and we have to get clearance from both before we sign a biotech alliance. We have a licensing department, and that has to be involved as well.

Compare with this description by the scientist-founder of a biotechnology company:

Ours is a young company, the average age is around 32, and very energetic. We don't have a corporate staff; senior people report directly to the president. We have a very independent, open attitude. The reason is that the closer one is to the academic model of research, the more difficult it is to organize precisely.

Finally, listen to how pharmaceutical executives describe some of the scientists from biotechnology companies who present their work as they seek funding:

. . . Most of the biotech lead scientists are arrogant. They can find work in a minute if their company fails—they are always head-hunted.

. . . All a biotech scientist knows is that this is the "first inhibitor for enzyme X, and why are you wasting my time asking business questions?"

In the biotechnology company, on the other hand, the founding or lead scientist may be described as follows:

The lore here is that [scientist X] is the creative genius. And, he's certainly treated like the Oracle.

CLOSING THOUGHTS

These few excerpts reflect the business model on which the pharmaceutical sector operates, and they also reflect the academic model on which the biotechnology sector operates. The business model supports "a careful strategy of asset redeployment." The academic model supports a "vision of what science can provide." The business model supports "therapeutic area teams with representatives from strategic marketing and medical." The academic model supports not "organizing precisely." The business model supports asking: "Are the numbers OK?" The academic model supports starting "with dreams."

The differences in values, beliefs, premises, and expectations generally held by people in biotechnology and pharmaceutical companies are consistent with the respective sector histories. We contend that the marked differences between pharmaceutical and biotechnology companies make sense, if we look to the past. Certainly, organization and discipline differences are important and complicating, because they also influence the perceptions and behaviors of people involved in alliances. The point we want to emphasize here is that the *differences in sector cultures are fundamental and underlie organization and discipline cultures.* Attempting to lead an alliance without understanding and accepting these sector differences is, we believe, courting failure.

Think differently . . .
   Lead differently . . .
      Make alliances work.

CHAPTER 4

PARTNER DIFFERENCES AND DISPARITIES

The contrast in cultures between the biotechnology and pharmaceutical sectors is not the only factor complicating alliance relationships. We believe that the large and unavoidable differences between biotechnology and pharmaceutical companies—in terms of physical size, financial situation, and partnership experience—also have a major influence. In this chapter we examine differences between typical biotechnology and pharmaceutical companies that are large enough to complicate the relationship and perturb the collaboration dynamics.

AN ELEPHANT AND A MOUSE

The contrasting origins and evolutionary trajectories of the pharmaceutical and biotechnology sectors have produced contrasting sector cultures. From our study of alliances, we conclude that the lenses through which people in each sector view the world is in some part determined by the respective culture. In addition to contrasting sector cultures, we find numerous other differences between the partner types, including some so large as to be quite out of proportion. We refer to these marked differences as *disparities*.

When alliances partners have disparities along a number of dimensions, a good metaphor is the circus ring with an elephant and a mouse act. The

relationship between partners can be described using such adjectives as "asymmetric" and "lopsided." In the Lucida–Pharma Sciences alliance, there were *not*, as one biotechnology executive noted, "two elephants in the ring." Although respected for their scientists and for their patents, Lucida Biotech was only a "mouse" in comparison with the magnitude of its partner. Pharma Sciences had revenues in the billions; Lucida worked on decreasing its losses. Pharma Sciences employed tens of thousands worldwide; Lucida employed tens of scientists locally. Pharma Sciences was engaged in about 20 biotechnology alliances at any one time; Lucida was engaged in one at a time. . . .

We have come to appreciate that effective collaboration does require a coordinated effort between an elephant and a mouse, rather than between two elephants, or between two mice (i.e., between equals). Significant disparities between companies in these sectors are unavoidable; thus, the impact of asymmetric relationships on the collaboration dynamics must be considered and accommodated.

In this chapter we identify some of the important general differences and disparities between pharmaceutical and biotechnology firms. We begin with a brief overview of industry-level information relevant to biotechnology alliances, to set the context of collaboration. Then, we discuss differences and disparities between the two types of firms and their potential impact.

THE CONTEXT OF COLLABORATION

The structure of the biomedical industry—a few very large firms dominating thousands of small ones—has been stable for decades. By the end of the 1990s, rising worldwide consumption of medicines coupled with mergers and acquisitions left this structure unchanged but greatly increased the scale of the top firms. It is now common for a major pharmaceutical company to have revenues of about $10 billion and to spend nearly 20% of those revenues on R&D. In contrast to these major league firms, it takes the combined revenues of all of the U.S. public biotechnology companies (around 300) to approach the revenue of one large pharmaceutical company.

Even in the simplest case, collaboration is never easy. Even if everyone is employed by the same organization, the desired lively, challenging, creative interaction among members of a team is the result of hard work, sophisticated interpersonal skills, and patience. An eight-person team consists of eight different personalities, experiences, backgrounds, training, perspectives, and so on. Crafting these differences into a unitary heart–mind–spirit (i.e., crafting an effective *team*) is an achievement.

In the case of biotechnology alliances, team members come from different companies in different sectors. They come from different disciplines and, with their different background and experience, hold different views of the world. Biotechnology alliances occur in a environment of rapid scientific and technological change (resulting in high uncertainty and ambiguity), are relatively short-lived, involve organizations of vastly different size, and are of quite different importance to each firm.

As we listened to our interviewees, we were struck by the realization that communication and interaction between scientists and executives from each type of firm appear to take place across a sizable power and dependence gradient. For us, the language we heard spoken by alliance participants is, truly, a language of asymmetry. We contend that thinking differently about alliances requires appreciating the unavoidable differences and disparities that create imbalance. Leading differently requires accommodating the differences and preparing for their consequences in terms of alliance dynamics.

DIFFERENCES AND DISPARITIES BETWEEN FIRMS

We want to make it clear to the reader that what may be only a *difference* in one alliance may be a *disparity* in another. We highlight the disparities in this discussion, because such marked differences can overwhelmingly complicate the lives of alliance scientists, managers, and leaders. The most obvious disparities are those of scale—especially, as the earlier discussion indicated, the relative size of revenues, R&D expenditures, employees and functional resources, and number of alliances in which each type of firm is involved. Other differences are measures of performance and organizational structure. In the following sections we describe the hypothetical "average" firm.

Scale

Revenues. Pharmaceutical revenues are measured in billions of (U.S.) dollars; biotechnology revenues, if any, are measured in millions. Like Lucida, most biotechnology firms have little or no revenue from sales. Their financial resources are derived from venture capital and, to a much greater extent, royalties and licensing fees from pharmaceutical companies. What is *cost* to the pharmaceutical company (licensing fee) is *revenue* to the biotechnology firm. What is a very modest portion of R&D resources to the pharmaceutical firm (royalties and licensing fees) is precious "fuel" consumed more or less quickly according to the "burn rate" of the biotechnology company. When Lucida's alliance was terminated and the fees from Pharma Sciences stopped, Lucida's burn rate was such that half their staff had to be laid off almost immediately. There was no noticeable impact on Pharma Sciences.

R&D Expenditures. Pharmaceutical R&D expenditures are also measured in billions (U.S. dollars), in contrast to the millions spent by

R & D as a Percentage of Revenue

85%   15%

Biotechnology   Pharmaceutical
Revenue   Revenue

Pharmaceutical Company
People Assets

Biotechnology Company
People Assets

biotechnology companies. However, if we examine the amount spent on R&D as a percentage of revenue, it accounts for about 20% of pharmaceutical revenues but an average of 85% of biotechnology revenues.[1] Moreover, the amount invested in research or discovery accounts for a much higher proportion of total R&D funds in a biotechnology company (as much as 100%). In a pharmaceutical company, research accounts for about one-third of the total for research plus development (i.e., R&D).

Employees and Functional Resources. A pharmaceutical company employs tens of thousands of people globally, whereas a biotechnology company employs hundreds locally. As a result, there is clear disparity between firms in terms of functional resources such as regulatory affairs, quality assurance, and so on. Pharmaceutical "people assets" are measured in "departments"; a biotechnology company may or may not have a single individual with expertise.

[1] Ernst & Young. *Biotechnology Industry Annual Report*, 1998.

In our case study, Lucida initially employed six senior scientists who constituted the management of the company with Rosenbloom, the founder. Before Santoro was hired as vice president of research, there was no research department. All scientists reported to Rosenbloom. In contrast, we know that Pharma Sciences had offices globally and employed tens of thousands of people. The company was so large that Dean, the CEO, could form "productivity committees" and implement a quantitative approach to portfolio management of R&D projects.

One biotechnology executive commented about the comparison between types of firms:

> There are tremendous resources at every level in the big company. Pharmaceutical companies have so many people in regulatory, or manufacturing— some of the best people in their profession. When we, on the other hand, had to put together a document for the FDA, we had no experience, and I had to bring in a couple of people who did. But, when that project ended, they had to be let go.

Alliance Numbers. Pharmaceutical companies may be involved in 20, 30, or 40 alliances at one time; biotechnology companies, one or a few. But, as a proportion of identifiable research programs or projects, the scale reverses. Alliances account for a small proportion of the pharmaceutical portfolio. They often constitute *the* portfolio in a biotechnology firm. Pharma Sciences was involved in 20 or so alliances with biotechnology companies at the time of its alliance with Lucida. In its entire history, Lucida had entered into only two alliances with large firms. Their other revenues came from contract work (small, one–off projects).

Performance Measures and Structure

Performance Measures. A pharmaceutical company is judged by sales growth, profitability, return on equity, and so on. A biotechnology company is judged by scientific news, good and bad, as proxy for quantitative measures. Their stock price volatility reflects this. When the alliance with Pharma Sciences was terminated, Lucida's share price "dropped precipitously by more than one-third." There was literally no change in the price of a Pharma Sciences share.

Organizational Structure. In general, large companies are organized differently from small ones. A pharmaceutical company is highly differentiated; there are many departments and divisions with relatively impermeable boundaries ("silos"). There is a hierarchy of multiple levels (the structure is very vertical), with formal promotion plans, rules, and performance criteria governing each level. A company like Pharma Sciences will utilize standard operating procedures and documented policies and will have well-defined systems for making decisions, such as a strict, quantitative approach to portfolio management.

A biotechnology company will have few departments and relatively few people, so that boundaries between functions will be more permeable. Some of our interviewees described this as "wearing a number of hats." In other words, one person will have responsibility across what would be considered separate functions in a pharmaceutical company.

A biotechnology company structure is more lateral than vertical (the oxymoronic "flat hierarchy"). Company procedures are unlikely to be formally codified in documents but are more likely to be kept in individuals' memories. Instead of a system for making decisions, the small organization will rely on ad hoc meetings and push for consensus. Particularly in the early history of the company, the situation may resemble the following:

> With power and financial interests so spread out, there was no choice but to try to run the company by consensus. Scientists did not know anything about running a business, however. There was no coherent business strategy and no criteria for picking and choosing among myriad possibilities (founder referring to his first biotechnology company).

Strategy

In addition to the fundamental issue of having a strategy or not (i.e., responding "totally opportunistically"), there are important differences between types of firms in several aspects of strategic decision making. These include what drives the decision to enter an alliance; the role of the alliance in company strategy; the worth of the alliance as perceived by senior management; the impact of "bad news" (i.e., scientific setbacks) on each partner; and each partner's strategic approach to risk.

Alliance Drivers. Senior management in the hypothetical average biotechnology company enters alliances because resources are needed to support current research. Senior management of the hypothetical average pharmaceutical company enters alliances because they expect long-term future profits from the technologies and/or products that may emerge years later. Thus, the alliance driver for biotechnology management is immediate funds, whereas the alliance driver for pharmaceutical management is potential future profits. Moreover, biotechnology management understands that simply having a contract with big pharma can be used as leverage for seeking additional research funds from capital and venture markets. Immediate money makes money immediately, so to speak.

Role of the Alliance. Given these drivers, it is not surprising that the role of a particular alliance in company strategy is crucial, vital, and acutely important for the biotechnology company (to "slow the burn rate") but modest (even slight) for the pharmaceutical company. In describing the alliance between Lucida and Pharma Sciences, Pitchly said:

> I worry that Pharma's priorities might change. What happens if another of their initiatives takes off? We're just one of many [alliances] to them, some number in the large queue of projects needing resources.

Compare the above perspective with this observation by a senior executive in a pharmaceutical company:

> We don't need to be successful with biotechnology alliances to be successful as a firm—but they need to be successful with them in order to be successful at all.

Remember: Although there is a difference in alliance drivers, leaders of both types of companies hope to derive a strategic benefit from the alliance.

Worth of the Alliance. *"You start with dreams. That's what we sell in this business"* (biotechnology company executive).

What is a dream worth? How do you determine the monetary value of a dream? In the hypothetical average biotechnology company, the value is very high. What has emerged from discovery may in fact be unique and eligible (as in Lucida) for speculative patents. But, pharmaceutical senior management appreciates the high failure rate of dreams. And, their calcu-

You start with dreams . . .

lation of worth will take into account not only that "this is the first inhibitor for enzyme [X]" but also that most "firsts" do not succeed in reaching the market. Because of the high risk of failure, pharmaceutical management is likely to assign a much lower monetary value to the alliance than is expected by biotechnology management. In determining worth, pharmaceutical management will ask such questions as:

> Is there clear business logic? Can we sell this? Will managed care buy this? Are the numbers OK?

News Impact. All scientific efforts are beset by problems, but news reports of problems have a different impact on biotechnology and pharmaceutical companies. A scientific setback can be life-threatening to the biotechnology company. Market capitalization, and with it the ability to generate more money to fund research, can plummet when a biotechnology firm is associated with bad news. Attraction to other pharmaceutical companies as an alliance partner also diminishes.

A biotechnology executive captured the difference in impact this way:

We take a beating when our partner says anything negative about the program. If they don't do the best possible job of presenting data, we'll suffer much more than they will. As an investment, they're evaluated on sales; we're evaluated on science.

If our partner discovers something about the quality of our data and says something publicly, we're affected. And the next time, even if the data are positive, there will be in the back of the investment community's mind something negative about us. We suffer this regularly.

Approach to Risk. Typically, there will be notable differences in each partner's strategic approach to risk. Pharmaceutical management diversifies risk by the breadth of their R&D portfolios, constantly balancing reward, risk, and current scientific data. As we heard in our interviews, biotechnology management usually has to "bet the farm" on one or two programs and often hang on "sometimes just by our fingernails" through the vicissitudes of a collaborative effort. The approach to risk of the hypothetical average pharmaceutical company is likely to be conservative. Pharma Sciences was known as "one of the most fiscally conservative of the majors," but their conservatism was not out of line with their peers.

SOME CONSEQUENCES FOR ALLIANCE DYNAMICS

Even this simplified overview of disparities and differences underscores how complicated the interorganizational context of biotechnology alliances can be. The differences and disparities are unavoidable. The resulting tensions cannot be eliminated, but their impact on the collaboration must be considered and accommodated. Perhaps the most sizable impact arises from relative differences in power, dependence, and influence. In the concluding section of this chapter, we discuss some of the consequences of differences and disparities—the issues and tensions—for alliance dynamics.

Power, Dependence, Influence

In an alliance of equals (two *elephants*, or two *mice*), each partner has sources of power and, therefore, the potential to influence some decisions. In an alliance of unequals—between an elephant and a mouse—tensions

Alliance of Equals   Alliance of Unequals

arising from imbalances of power, dependence, and influence will be high. The first imbalance that we will address is that of power deriving from relative magnitude of resources.

Power. The pharmaceutical company is a "powerful buyer" of biotechnology expertise in the economic sense of the term. Even though he was not aware of the formal definition, a biotechnology executive we interviewed recognized this, when he said: "It's a buyer's market!" As defined by economists, a buyer is powerful if (among other attributes)[2]:

- *It is concentrated or purchases large volumes relative to seller sales.* As our discussion of the structure of the biomedical industry showed, the pharmaceutical sector is concentrated: A few large firms account for a sizable proportion of sales and R&D expenditures. In addition, the hypothetical average pharmaceutical company purchases large volumes of biotechnology research via strategic alliances, compared with the sales of an individual biotechnology firm. The alliance portfolio of one pharmaceutical company may exceed the sales of tens or even hundreds of small biotechnology firms.
- *It faces few switching costs.* A pharmaceutical company can and does enter alliances with numerous biotechnology firms. If one project fails, the company can switch (change) easily to another partner and is not locked in to one firm. The "sudden wealth of targets and compound diversity" suggests that a number of biotechnology companies may provide similar opportunities for one pharmaceutical firm.

[2]Porter, M. *Competitive Strategy: Techniques for Analyzing Industries and Competitors*, New York: The Free Press, 1998.

- *The industry's product is unimportant to the quality of the buyers' products or services.* For the end consumers of a drug—physician and patient—a major determinant of quality is the brand name of the pharmaceutical company selling it. The biotechnology industry's product, intellectual property that is incorporated into the therapy, may be strategically vital for the pharmaceutical firm but is invisible (and unimportant) for these buyers by the time the product reaches the market.

For biotechnology executives and scientists, the powerful buyer issue looms large. In the words of one of our interviewees: *"They pay; they say. . . ."*

Dependence. The second imbalance we will address is that of dependence, which is related to power and which results in additional tensions around influence. The potential ability of one company to influence another depends not only on the relative power of that firm but also on how dependent the firm is on the other's resources. Usually, the more powerful firm has resources the other needs and is in control of their allocation, especially when there are few, if any, alternatives. If there are alternatives (whether they are being used or not), the power of the organization holding the resources is diminished.

If we define dependence as the relative level of resource investment of each partner, then it is clear that on the pharmaceutical company side the investment is low; on the biotechnology company side, it is high. Low investment relationships imply a low degree of dependence; high investment relationships imply a high degree of dependence.[3] At an industry level, as we noted earlier, alliance revenue now accounts for more than half of all capital invested in biotechnology firms. The average biotechnology company is highly dependent on a its pharmaceutical partner(s). As the experience of Lucida indicates, decisions of the pharmaceutical partner can influence a biotechnology firm's survival.

The Influence Seesaw. In an alliance, the pharmaceutical partner "holds" monetary and functional resources that the biotechnology partner needs. The alternatives, private equity placements and public stock offerings, may be out of reach for most small firms. Thus, the pharmaceutical

---

[3]Auster, E. The Interorganizational Environment: Network Theory, Tools, and Applications, in Williams and Gibson (Eds.), *Technology Transfer: A Communication Perspective*, Newbury Park, CA: Sage Publications, 1990.

partner is powerful over the duration of an alliance and can exert more influence on the decision-making processes. Whether this powerful partner actually "flexes his muscles" to influence decisions is contingent on two conditions. First, if the alliance is of very minor importance to the pharmaceutical company, senior management may allow the biotechnology partner to have more say. Second, if the biotechnology partner is able to convince the more powerful pharmaceutical company of its own areas of competence, biotechnology scientists and executives may be able to influence decisions in those specific domains.

In addition to monetary and functional resources, the partner who is better able to access and process information critical to the success of the alliance may gain some power and ability to influence the other. The fluid and complex dynamics in relationships come about because neither partner wholly controls the information and decision processes, and neither partner has authority that is seen as legitimate by all concerned. Power as a factor in biotechnology alliances is not static. It may change as problems and situations change, as relational dynamics change, and as other options become available. The power and dependence dynamics of the alliance must be examined at various points in time, and with respect to particular decisions. In addition, although an organization may be perceived to have various sources of power, it is individuals and groups who make decisions. As decision processes unfold, individual power and influence within the alliance may shift, as coalitions are formed and reformed.

Finally, for the less powerful partner in a relationship, the objective may not be to accrue more power but to use existing power in a more effective way. Managers and scientists in a biotechnology company must "make things happen" and influence the course of the alliance, without benefit of formal power/authority. When describing his approach to getting things done in an alliance with a tough-minded pharmaceutical partner, one biotechnology executive commented: "It's all about the power of personality."

Below, we address the less sizable but still important issues and tensions that arise from other differences between pharmaceutical and biotechnology firms.

Resource Issues and Tensions

Cost Versus Revenue. Money to fund an alliance is accounted for as operational cost by the pharmaceutical firm but as revenue by the biotechnology firm. The resulting impact on collaboration dynamics may be inferred from the natural vocabulary associated with these terms: "Costs" must be "reduced"; "revenues" must be "increased." But, cost-reducing and revenue-increasing activities result in countervailing forces on the alliance. Conducting additional tests to confirm scientific data costs money. The pharmaceutical firm is likely to request more data, while the biotechnology company will want to save money. Collaborating scientists will feel (implicit or explicit) pressure to reduce costs from the pharmaceutical side and to increase revenues from the biotechnology side. Scientists will always feel those opposing forces; project managers will feel them acutely.

Structure (or Not). At the leading edge of science, with no or little revenue to fall back on, biotechnology companies cannot afford to have people spend energy and effort on creating administrative and managerial systems, policies, and procedures. To a certain extent, biotechnology companies are characterized by a lack of structure. Large pharmaceutical companies have income streams deriving from previous successes. Money and time have been spent to compose training manuals, to design orientation programs, and to set up sophisticated communication and intelligence systems.

In an alliance, there will be tensions arising from decisions on what aspects of the pharmaceutical structure should be replicated in the collaboration. Some individuals will be comfortable with (and want) more structure, formal processes, and decision systems; some will want to operate more loosely. We want to make clear that there is no one right answer in this regard. What is important is that the lived structure facilitate effective collaboration.

Climate of Plenty (or Climate of Lack). The hypothetical average pharmaceutical company operates in a climate of plenty. There are sufficient monetary and functional resources, and they are relatively easy to muster. The hypothetical average biotechnology firm operates in a climate of lack. Discretionary resources are scarce and not readily available. The company lacks sufficient money and never employs enough people. Tensions are likely to arise if the pharmaceutical partner fails to appreciate the scarcity of resources and has unrealistic expectations, particularly in terms of project time and scope, of the biotechnology firm.

Strategic Issues and Tensions

In an ongoing alliance, tension may arise because company objectives are very different, and each partner values its own tactics for achieving success more than it values their partner's tactics. Tension may also result from ignoring or overlooking strategic differences after the deal-making stage. Leaders should never assume that such differences effectively disappear, once there is an agreement in place.

At minimum, the details of strategy will be different, and senior managers from both partners may also have very different expectations of the alliance *per se*. Often, at the start of an alliance, agreement on the premise, goals, and value of the collaboration to each party may be assumed but rarely assessed. Also overlooked is the need for clear definitions and clear agreement on the scope of the collaboration (what is, and what is not, included in the terms of the contract). Conversations about strategy should continue as the alliance forms and develops and should include more than just the senior people.

The first milestone is a key marker and represents a good point in time to revisit and reevaluate the terms and time frame of the contract. As a result of efforts to achieve the first milestone, there may be more clarity about possible outcomes. For the ease of managing toward success, greater explicit clarity is to be desired at this point. Hence, reevaluation may be strategically prudent as well as beneficial.

A less-than-specific outcome (at the start of an alliance) does not preclude clear and specific milestone definition; in fact, it may demand it. A senior manager in a pharmaceutical company stated the following:

> In the basic, fundamental research kind of alliance, it is not clear what the product will be. But, in the agreement between us there must be clear milestones that everyone agrees will represent progress. If there are not, that opens the door for conflict. There may be milestones, but if they are not clear, distinct, and sharp enough to avoid disagreement as to how progress is being measured, then they are not helpful.

An alliance is also likely to have different degrees of importance in and relevance to each strategy. The salience of the alliance to senior managers, along with the attention the collaboration receives, will reflect the importance and relevance the alliance has in the context of corporate strategy. Although a particular alliance may be crucial for the biotechnology company but only modestly important for the pharmaceutical partner, progress in achieving the outcome of the collaboration (i.e., meeting the milestones) may raise the role, importance, relevance, and salience for pharmaceutical management:

> When the product looks as if it might be really good, our senior managers get nervous. And, the better the product looks and the further downstream in the process, the more our people want to become involved.

The desire to grab control is not just in development. Senior managers want to grab control as they hear about the product and the competitive situation. So, there is "grabbing control" throughout our organization over the course of the collaboration, depending on how good the product appears to be (pharmaceutical company vice president).

CLOSING THOUGHTS

Of central importance to us in this book are the leadership tasks associated with creating new knowledge in the interstitial space between two organizations. These tasks include, for example, focusing the collective effort of two disparate organizations, managing individuals, and establishing and maintaining collaborative relationships under what might be termed "hardship" conditions.

The challenge is to make alliances work in spite of differences in sector and organizational cultures, asymmetries of power and influence, and diversity of backgrounds. Disparities and differences are part of the reality of alliances between biotechnology and pharmaceutical companies. Research collaborations are a coordinated effort between a mouse and an elephant, rather than a marriage of equals. In this situation we argue that the "mouse" should bear a greater responsibility for the effectiveness of the alliance dynamics.

Think differently . . .
> Lead differently . . .
>> Make alliances work.

PART III

LAYING THE GROUNDWORK

To this point in the book, we have focused on background and context issues as they apply to biotechnology alliances and their leadership. In this part, we move from the more general to the more specific, focusing on the preparation essential to alliance effectiveness (and the desired alliance success).

In Chapter 5 we address what a biotechnology organization should have in place before any collaboration begins, and we identify the critical leadership roles. In Chapter 6 we look at what people in the biotechnology company should know/learn about the specific pharmaceutical partner. In Chapter 7 we discuss issues relevant to preparing for the first meetings, both the informal meetings and the formal kick-off event.

We assume that an alliance will take place and that a partner has already been selected, for three reasons. First, research alliances are the lifeblood for most biotechnology companies. Most of them do not have the luxury of deciding whether to partner with a larger pharmaceutical company. If they can, they will, so they should be ready. Second, the reality is that choosing a partner is typically the province of the larger company. Few biotechnology companies are in a position to choose from among several possible partners. Third, as our focus is on making alliances work, we can assume that the deal has been struck. The leadership task is to make it work, regardless. . . .

Before we address the more specific issues relevant to making alliances work, we need to describe two concepts that will be very important to our discussions in the remaining chapters of this book. These are the concepts of *uncertainty* and *equivocality*.[1]

If everything were perfectly known about a subject (e.g., the partner firm, the scientific experiments), there would be no *uncertainty*. Uncertainty means that some knowledge is lacking, that the information at hand is imperfect. Uncertainty refers primarily to the "what" and "how to" aspects of the collaboration, including aspects of the partner.

Equivocality refers to multiple and, sometimes, competing interpretations about some thing. Equivocality (ambiguity) exists when:

- Issues are unclear, as opposed to unknown.
- It is difficult even to articulate the right questions.
- There is confusion because different people interpret the same information in different ways.
- The "language" being used is not understood and/or interpreted differently by people in different functions (e.g., R&D versus strategic marketing).

Whereas the level of uncertainty is reduced through information processing activities, the level of equivocality is reduced through transactive (two-way) communication that creates shared meaning and understanding. Simply disseminating information does not reduce ambiguity.

It is possible to judge the quality of the result of reducing uncertainty as well as the process by which it is reduced. We can ask whether the best information sources were used, we can evaluate the completeness of the information provided, and so forth. Under conditions of equivocality, however, whether the "best" solution is attained is not particularly relevant (or even answerable). What matters is the extent to which people come to "see" things in the same way and agree to a common solution, decision, or interpretation (meaning) they attach to the data and language used.

For alliance leaders, understanding these differences and being able to respond appropriately are crucial to effective collaboration.

---

[1]We owe a sizable intellectual debt to J. Galbraith, whose book on complex organizations (1973) described uncertainty and its impact on the structure of work. We also acknowledge our debt to early writings of R. Daft and R. Lengel, who addressed equivocality and communication in organizations. These authors are cited in the next chapters.

CHAPTER 5

PREPARING THE ORGANIZATION

If biotechnology alliances will play an increasingly important role in the biomedical industry, then preparing for effective collaboration makes good sense. As one researcher commented to us:

> I would venture that we don't think enough about collaboration ahead of time. So, our expectations get distorted. When the going gets rough, when we have to cope with problems that inevitably arise, we're not prepared.

Effective collaboration, whether or not the outcome is commercially successful, does not just happen. Thinking "about collaboration ahead of time" involves, for example, considering the following:

- How the alliance fits with each partner's strategy (or vision)
- The commitment of senior executives in each firm
- The availability of scientific teams with appropriate expertise and work norms that will support collaboration
- Whether organizational structures and systems will facilitate and reinforce collaboration or may impede it
- Some implications of both sector and organizational cultures—values, beliefs, norms—and how to ensure these will encourage creativity, achievement, and collaboration (a difficult mix, assuredly).

Checklist for Effective Collaboration

✓	Does the alliance fit each partner's strategy?
✓	Is there leadership commitment?
✓	Are the right people on board (appropriate technical skills, good work norms)?
✓	Are the right organizational and structures and systems in place?
✓	What are some implications of culture differences for the alliance?

Leaders on both sides of an alliance are responsible for laying the groundwork within their own organizations, well before an alliance begins. To a large extent, effective collaboration depends on leaders' ability to facilitate the following:

- Development and maintenance of collegial, professional relationships that will sustain the alliance over its duration
- Smooth flow of information between partners
- Processes that will result in shared understanding and agreement between partners
- Individual and collective learning, using current and prior alliance experience(s), to improve ongoing and future partnering.

Preparation begins before the alliance starts, and institutionalizing the learning comes after the alliance ends. Certain tasks (and the roles of certain people) will be more or less prominent, waxing and waning over the course of the alliance, depending on what issues and problems are at the fore at any given time. Responsibility for becoming a good or better partner, however, never goes away.

Before we address the subject of preparation, we want to clarify the difference between alliance success and alliance effectiveness. We are not concerned with commercial success. What occurs in the market is outside the scope of this book and is downstream from the collaboration that produced the innovation. We are concerned with alliance success, which we define as achieving the scientific and technological progress stipulated in the formal contract and other alliance documents.

More importantly, we are concerned with effectiveness, which we define as an alliance that works well. Working well includes the obvious characteristics of good collaboration, communication, and commitment, but it is more than that. An effective alliance also supports and enhances the intellectual and skill competencies of partner firms, provides multiple opportunities for learning, and is judged by leaders on both sides to be valuable and worthwhile. *Ensuring alliance effectiveness is at the heart of this book.*

We believe that alliance success is more likely under conditions of alliance effectiveness; in other words, that troubled alliances are less likely to meet their stated scientific and technological objectives. For this reason, we focus on the organizational and role issues critical to alliance effectiveness. These issues we can address, because these are well inside our field of competence.

This chapter addresses the preparation that must be undertaken within the biotechnology organization. In the next chapter, we will delineate what biotechnology leaders should consider in the course of the interpersonal and organizational due diligence efforts that must precede any alliance. For now, we describe how biotechnology leaders should prepare their own "house." (Some of the discussion, of course, also applies to preparation within the pharmaceutical company, especially for those companies whose leaders wish to become better partners and develop a good alliance reputation.)

In the next section we briefly review the fundamental issues of strategic fit and leadership commitment. Then, we describe issues and actions relevant to alliance leadership roles and who should fill them. In the concluding section, we describe issues and actions relevant to the organization, particularly structure, systems, and culture.

Alliance effectiveness: good process

Alliance success: good outcome

FUNDAMENTAL ISSUES

Strategic Fit

For most, if not all, small biotechnology firms, *strategy*—formally articulated, codified, revisited, and revised on a regular basis, and so on—is a foreign concept, a luxury attainable only when a firm is commercially successful. We have come to appreciate that the large partner will have an articulated strategy, but the small one usually will not.

For example, when pharmaceutical executives and scientists discussed alliances with us, they talked about overall corporate strategy and how research programs and projects (including the external alliances that were part of the portfolio) supported that. They explained where and why they focused their R&D resources:

> We pursue a careful strategy of asset redeployment, and we target alliances that are in selective therapeutic areas where we have a strong franchise.

In contrast, when we listened to executives and scientists from biotechnology companies, we rarely heard the term "strategy" without prompting. Consider how a Lucida senior scientist described their strategy as *organ of the month club*. Other responses by biotechnology executives and

A good strategic fit.

scientists to our questions about strategy highlight this small partner/big partner contrast:

> . . . Strategy? It's totally opportunistic.

> . . . We try to deliberate our options, but we end up taking the best economic terms we can get.

> . . . Our strategy? To slow the burn rate and partner the risks . . .

However, even if there is no formally codified strategy, we contend that an alliance must at least be consistent with the trajectory described by the biotechnology company's long-term vision of what it hopes to achieve. As a biotechnology executive said: "You start with dreams. . . . We have a vision of what science can provide and what that might mean to investors." From the perspective of the biotechnology company, we define fit as an alliance that

- Is consistent with the vision of what their science can provide
- Can make good use of the intellectual and skill competencies in the organization
- Provides opportunities for learning consistent with the vision
- Is expected to add demonstrable, "vision-consistent" value to the company, regardless of the commercial success of the outcome.

If the alliance will only "slow the burn rate and partner the risks," if it is wholly "opportunistic," then competencies and value will be added randomly, if at all. (Few visions are ever achieved wholly by chance.) We are not minimizing the real pressures on biotechnology executives and scientists to ensure survival of the company. However, we emphasize that *survival* and *fit* are not mutually exclusive. If the situation is "partner or perish," then everyone within the company should work to preserve some resources, intellectual as well as monetary, for future alliances that will help in the attainment of their vision.

Leadership Commitment

Following on from our discussion of strategy and its lack, this topic may appear ironic to readers for whom alliances are a necessity rather than a choice. From the biotechnology perspective, what does it mean to be committed to a necessity?

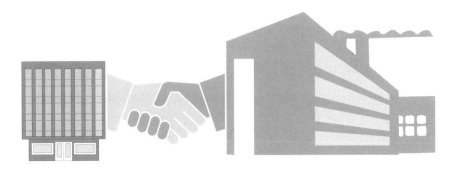

Leadership commitment.

In fact, we separate the necessity for alliances (as an economic fact of life in the biotechnology sector) from the necessity for commitment to the collaboration. Senior management of both the small and the large partners must be committed to a specific partnership. Certainly, the commitment of senior managers was clear at Lucida and Pharma Sciences. As Lucida's VP of strategy said: "Alliances work well if the folks at the top are personally dedicated to it. . . . The fact that you have commitment at the top changes the interactions at the lower levels."

But, people must also be committed to the value of collaboration, which is not as simple as it sounds. If everyone is committed to this value, to being collaborative, they are committed to being flexible and doing what is required to make the collaboration successful, even if they have to do something different from what is customary.

Being committed to this value also entails what one of our interviewees described as a commitment to "allow for failure, tolerate and encourage risk-taking, and handle failure well." Just because the biotechnology sector is a more risky sector does not mean that biotechnology leaders, by definition, encourage risk taking (or allow for failure and handle failure well). However, we want to make it clear that "risk" and "failure" must be understood in the context of the decision to partner *because* the alliance is consistent with the company's long-term vision. In this context, economic risk is reduced. Such an alliance supports and enhances the firm's competencies and provides multiple opportunities for learning. The latter will add demonstrable value to the company, even if the outcome is a commercial failure.

If commitment to the value of collaboration is strong, then the alliance will benefit in a number of ways. One scientist admitted that strong commitment "helped us get over a lot of the blood and guts that came out at various stages of the alliance and overcame a whole lot of interpersonal problems that developed." Everyone must also be committed to a set of principles by which collaboration with external partners will be judged. After a negative alliance experience ("the worst agreement that we ever went into"), a senior biotechnology executive described the lesson they learned as follows:

> It made people begin to think about what kind of research we were really going to conduct, and what principles were important to preserve. We had a knock-down, drag-out fight within the group, but we reached a set of principles on which we built all our future agreements.

Ideally, both partners will gain from the complementary contributions of each other's expertise. For a biotechnology firm, a central principle to preserve is learning:

> There is no point in entering a partnership with a group that does not think, because then my own people don't think. Even though certain Pharma company teams are very arrogant, their people *think*. So, when our scientists interact with them, they have to be sharp. If we do not learn something, I find it hard to justify being in the partnership (biotechnology vice president).

LEADERSHIP ROLES

With the two fundamental issues of fit and commitment as context, the next issue we address is how to prepare the people in the company for effective collaboration. Our concern here is with the critical roles that must be filled. Although the individuals who will assume these roles in any particular alliance may not be known ahead of time, it is important that the kinds of roles required for an effective and successful alliance be recognized. We will also address the characteristics of "the right people" to fill these roles.

Vantage Point

We believe that there are four critical leadership roles, each with unique alliance responsibilities. Note that our discussion is about roles, not individuals. We do not define the specific actions of an individual; rather, we are taking a role approach and dealing with prescribed patterns of behavior. In many situations a single individual will, sometimes must, assume more than one leadership role (we will address this later in the section). The four roles, along with the vantage point or purview from which each deals with alliance issues and problems, are as follows:

Leadership Role	Vantage Point
Commitment Champion	5000–15,000 feet
Process Champion	500–7500 feet
Science and Technology Champion	500–7500 feet
Alliance Team Members	1–5000 feet

Three of the leadership roles we describe are championship roles; the fourth (team member) is a somewhat less evangelical role. Other than the Commitment Champion role, the remaining leadership roles will have specific responsibilities only in the context of a specific alliance.

Commitment Champion. At the senior levels of the biotechnology organization ("5000–15,000 feet"), the Commitment Champion has general and ongoing responsibility for overseeing alliance preparation, for participating in due diligence, and for encouraging top management commitment to an alliance over time. Whoever assumes this role must champion the alliance cause in general, as well as assess (and reassess) preparation, readiness, and commitment as the alliance experiences predictable "lumps" and "bumps" along the way.

Process Champion. The Process Champion has specific responsibility for keeping an alliance invigorated, focused, and functioning smoothly ("500–7500 feet"). In this role, the emphasis is on relationship dynamics: how scientists are working together within the team, and how they are working with scientists in the partner organization. Note that the vantage point of the Process Champion partly overlaps, so to speak, that of the Commitment Champion.

Science and Technology Champion. The Science and Technology Champion is, as the title implies, responsible for the scientific and technological progress of the alliance. Subject matter expertise is crucial for this role, and issues of quality and rigor come to the fore. But, the Science and

Technology Champion must also keep the dream alive, make certain that even small scientific and technologic achievements are recognized and appreciated, and ensure that inconclusive or negative results do not derail the scientific efforts. Because of the latter responsibilities, the vantage point of the Science and Technology Champion is the same as that of the Process Champion ("500–7500 feet").

Alliance Team Member. The Alliance Team Member is responsible for leading efforts in specific areas of expertise ("1–5000 feet"), which is why the vantage point partly overlaps that of the Science and Technology Champion. Scientists working as members of an alliance team may not perceive themselves as "leaders," but of course they do serve a leadership role in the advancement of the science and technology. For each alliance team member, leadership will be most evident in decisions about how to move forward, how to interpret results, what the next steps should be, and so on.

As the prior discussion should make clear, each role defines a somewhat different set or constellation of responsibilities. Below we highlight the unique aspects of the four leadership roles by providing the "mantra" that goes with each of them:

Leadership Role	Mantra
Commitment Champion	We believe.
Process Champion	We relate.
Science and Technology Champion	We succeed.
Alliance Team Member	We create.

The three champion roles do share responsibility regarding the kinds of interpersonal processes that promote collaboration and alliance effectiveness. In some ways, all champion roles are "head cheerleader" roles, as the following attributes suggest:

Alliance Champions . . .

Have passion
Articulate a common purpose
Model cooperation
Listen well
Are respectful

Elicit and encourage trust
Deal effectively with conflict

We will come back to these four roles in later chapters, as we examine alliance tasks over time. As a general rule, communication between and among the people taking on each of the roles is critical. Alliance effectiveness is enhanced by seamless role coordination, which requires open communication among the different role holders.

Roles and Persons

As the prior discussion illustrates, each role makes different demands on individuals. Thus, just as there are the right roles for effective collaboration, so too there are the right people to fill these roles. Here, we briefly describe general characteristics of the "right people," beginning with roles closest to the science and working our way to the executive level of the biotechnology company.

Alliance Team Members must have the appropriate expertise and an enthusiastic willingness to cooperate. The best team members not only are leaders in their areas but also are willing to work in less familiar territory. Only when that occurs can the team's efforts and successes be greater than the sum of its individual parts. Alliance Team Members should be focused on creating new knowledge, new models, new tools, and so on, and should understand the value of both individual and collective efforts.

Science and Technology Champions must be subject matter experts who truly understand the benefits of sharing and challenging data, information, and perspectives. The right people for this role have the self-confidence to demonstrate their competence and are willing to admit what they do not know. These champions must believe in doing whatever is needed to achieve scientific and technological success. This requires the dogged pursuit of high-quality, verifiable, reproducible, and well-documented experiments; it also requires the treating of negative results as a learning opportunity, not as failure.

Process Champions must be capable of facilitating and encouraging good relationships among different individuals who share a common purpose. The right people for this role enjoy dealing with the dynamics of collaboration among team members, across discipline boundaries, and across organizational boundaries. They must possess the competencies required to create and maintain productive dynamics. They must also understand the importance of developing and maintaining communication networks through which "news" and ideas flow easily. Finally, they must be able and willing to coach others in these skill areas.

Commitment Champions understand how to garner and maintain support in both alliance organizations. Within the biotechnology company, the Commitment Champion can ensure that the organization genuinely values being a good alliance partner. The right people are capable of looking "outward" and can assess and reassess commitment on the other side. Commitment Champions believe in the value of alliances generally and in the added value of particular alliances. Such belief will sustain both individuals and groups through the inevitable tough times.

The Right People for Combined Roles

Particularly in a small biotechnology company, a single individual might well be expected to assume more than one leadership role in an alliance. There are simply not enough people to go around. In our experience there are two potential combinations, but one is a better choice. The first merges the *Process Champion* and *Science and Technology Champion* roles. The second merges the *Commitment Champion* and *Process Champion* roles.

Process and Science and Technology Champion. This combination is *de facto* what happens in many companies and might seem to make sense, given the common vantage points (500–7500 feet). Unfortunately, this is rarely an effective combination. Why? One role is responsible for relationship *process*, and the other is responsible for scientific *progress*. The skills and talents needed to succeed in both are unlikely to be found in a single person. One scientist may be quite capable of leading the science yet incapable of developing communication networks and resolving conflicts effectively, and vice versa. (We mention the technical skills first, because the common assumption is that a technically skilled person is capable of leading a team. Infrequently, if ever, is a "people person" assumed to be a technical expert, without strong evidence to that effect.)

In Lucida Biotech, for example, the key bench scientist (Will O'Brien) was described as difficult, as valuing "science and creativity over personal interaction," and as "not a team player." One of his colleagues said he was

. . . very smart scientifically, but not managerially inclined! He believes that, if you're the boss, you tell people what to do and they do it. . . . [During] joint team meetings, O'Brien was publicly dysfunctional.

To management's credit, neither Pitchly nor Rosenbloom wanted to put O'Brien in charge of the alliance team. On the other hand, during the brief period that Pitchly led the collaboration, one of the senior scientists had to help with "stealth management" and "tell Pitchly what points people should address." The implication was that Pitchly was not, nor was he assumed to be, a technical expert. It may be feasible, and a good choice, to assign the Process Champion role to an outside consultant, if the only other option is a "Will O'Brien."

Commitment and Process Champion. Combining commitment and process leadership roles is the better solution, because (1) neither role requires a deep understanding of the science, yet (2) both roles require a good understanding of people and organizations. In this arrangement, the Commitment Champion might serve as consultant, or behind-the-scenes process mentor, to the Science and Technology Champion. The drawback to combining these roles is that the person who is drawn to executive-level issues and concerns may not appreciate the strategic value in what appear to be *only* everyday operations of the alliance team.

If roles must be combined, the choice should take into account the passions and interests of the people involved, as far as possible. If a person's heart is not in a particular alliance role, genuine leadership will not emerge. Assigning a championship role to someone who does not enthusiastically embrace that challenge does not work well. Although our advice cannot always be followed, the consequences of assigning roles to neutral, if not resistant, individuals should certainly be considered.

In whatever people-role resolutions a company chooses, communication among the people involved in filling these roles is critical. Alliance effectiveness is enhanced when there is relatively seamless role coordination.

STRUCTURE, SYSTEMS, CULTURE

In addition to ensuring what might be called "role readiness," alliance leaders must also ensure organizational readiness. Below, we address three organizational-level issues that are fundamental to effective collaboration and alliance effectiveness: structure, systems, and culture.

Structure

There has to be some sort of organizational structure that can support the *development* of collaboration. And, any time a collaboration is envisioned, someone in the organization should assess what is needed to make it a success, and not just in the initial phase (biotechnology company president).

At the end of the day, the structure of both organizations must be designed to support not just external collaboration but internal as well. The reason for that sweeping statement, as one of our interviewees noted, is that "modern molecular biology is a collaborative enterprise." A research structure that supports a silo mentality will impede internal and external scientific effectiveness. A structure that promotes wide-ranging, lateral discussion will support scientific effectiveness *and* good collaboration.

The "good news" is that preparing the organization structure for alliances is a win–win (external/internal) effort. The "bad news" is that organizing work (i.e., designing structure) is not accomplished on paper. Anyone can draw a novel graphic to depict structure, but that is what it will remain, a novel graphic. Structure cannot be legislated; structure is *lived*, consistently and continuously. That is what makes good structure difficult to achieve and what makes culture so important, as we will discuss later. Moreover, there are many ways to organize/structure work and not one best way. Finally, entire books are written about structure; we are only going to describe a few, key issues and actions in this section.

Structure Defined. Organization *structure* is often taken to be interchangeable with the organization *chart*, illustrating who reports to whom. Structure, as we use it in this chapter (and as implied by the earlier quotation referring to a "structure that can support the development of collaboration"), is a much richer concept. Fundamentally, organization structure is the pattern by which people relate to each other (which is not the same as report to each other) and communicate with each other. The organization chart portrays formal lines of authority, and those may be quite different from how people actually relate and communicate.

Pattern of Relating and Communicating (Direction). Although there are many ways of organizing work, not all are equally effective. There are two fundamental patterns or "directions" of relating and communicating

Lateral Relating and Communicating Structure

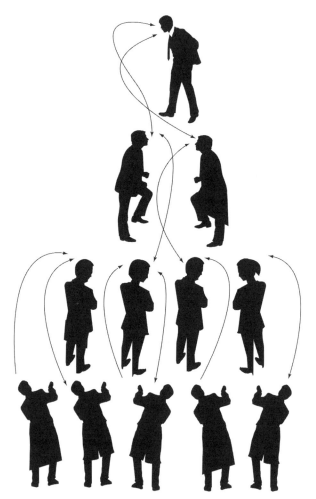

Vertical Relating and Communicating Structure

(all structures are permutations of these), but only one supports effective collaboration. The first direction is vertical, which reflects superior-to-subordinate, pyramidal, hierarchical patterns of relating and communicating. The second is lateral, which reflects equal-to-equal, peer, horizontal patterns of relating and communicating. *Lateral* is the required direction for effective scientific collaboration, because lateral communication supports equal-to-equal relating and open, candid, and informal relationships. Peer relationships and open, candid, and informal communication, in turn, ensure that there will be fruitful scientific debate and intellectual challenge among alliance team members. Information search processes will be wide and unconstrained by assumptions of authority; they will thus produce results surpassing processes that are narrow and constrained.

A vertical pattern is the consequence of (i.e., reflects) an assumption of superior–subordinate relationships modeled on parent and child. Included in this assumption is the notion that knowledge and authority reside "on top." The communication flow will, by definition, be primarily top-down (direction), with some bottom-up (feedback and information). This pattern contradicts what we know works best in R&D and, by extension, in an R&D alliance. Research must be organized laterally, both internally and externally. If an alliance is organized vertically, it will be characterized by (in the words of one of our interviewees) a "strong hierarchical structure that inhibits communication among the key players."

Perhaps surprisingly to some readers, "lateral" is not necessarily an attribute of small organizations, nor is "vertical" a necessary attribute of large ones. Our (very small) example company, Lucida Biotech, appears to have had troubled peer/lateral relationships, as illustrated by the luncheon meeting among Santoro (the new vice president of research) and three scientists. Santoro had observed "little camaraderie among the three scientists—it was as if they worked at three different companies." A clear sign of a vertical pattern/direction was revealed by Santoro's realization that Lucida "essentially revolved around the scientific founder, Rosenbloom." Typical of a vertical structure, authority was assumed to reside at the top ("*power* to the scientists meant access to Rosenbloom").

Functional Borders (Demarcation). Just as there are two fundamental directions of relating and communicating ("up/down," "side/side"), so too there are two fundamental types of borders or demarcation between groups, both internally and externally. We define these types as rigid and impenetrable, or flexible and permeable. Not surprisingly, only flexible and permeable demarcation of groups supports effective collaboration.

Flexibility and permeability (and all aspects of effective structure) cannot be legislated but have to be lived. Maintaining this type of demarcation requires continual modeling and oversight, because another attribute of modern molecular biology is the natural tendency for disciplines (especially new ones) to defend their territory:

> There are two problems I see limiting the degree of collaboration we can achieve. First is the departmental organization and the real difficulties that arise in trying to get people to work across disciplines. There is no easy mechanism for this, and there are enormous squabbles when people try to do cross-disciplinary research. Second, each discipline has its own territory and is fiercely defensive (consultant to biotechnology startups).

Living a Structure. One of our interviewees provided an excellent, one-sentence summary of the wrong organization structure:
He elaborated by saying:

> Any organizational structure that gives people the excuse to avoid collaboration, by reinforcing impediments to collaboration, is a mistake. You have to have a structure that reduces interpersonal and cultural impediments.

We want to conclude this discussion by identifying two signs of the right lived structure.

The first sign of the right-lived structure is minimum hierarchy (hierarchy presumes superior–subordinate roles and knowledge at the top). When everyone, including senior executives and chief scientific officers, admits candidly (both privately and in meetings) to what they do not understand, then hierarchy is minimized, lateral structure is created, and excuses are eliminated. When alliance leaders ensure that scientists in the team defer to the person with competence for the problem at hand, rather than to a title, they are minimizing hierarchy, and so on. When group leaders facilitate discussions in which every person's contribution is heard nonjudgmentally, they are minimizing hierarchy, and so on.

> *Any structure that gives people the excuse to avoid collaboration is a mistake. . . .*

The second sign of the right-lived structure is minimum discipline squabbles and territory defense (in other words, minimally rigid and impenetrable borders). Using interdisciplinary groups to develop criteria for internal scientific promotion, to set up journal clubs, or to institute a series of outside speakers in key subjects promotes permeability between disciplines. Another tactic is to promote people who value disciplines other than their own. When people frequently relate and communicate across disciplines and departments, they are much more likely to do so at the bench and in an alliance.

Bad signs (or, signs of bad lived structure) can be readily observed. If meetings are not candid, if intellectual competence is overlooked for another attribute, and if some people's comments are consistently ignored, the resulting structure "gives people the excuse to avoid collaboration, by reinforcing . . . impediments to collaboration."

Systems

The second issue we address to prepare the organization for effective collaboration is the set of systems that includes recruitment, performance appraisal and reward, communication, and decision-making. Even if a small biotechnology firm does not utilize codified or formal systems, it will rely on general practices, and we will use the terms *systems* and *practices* interchangeably. Like structure, general practices (systems) must support the development of collaboration, internally as well as externally.

Recruitment. "Modern molecular biology is a collaborative enterprise." If we believe that this is true, then it makes sense to recruit people who are willing to collaborate. That does not mean jeopardizing intellectual competence for collegiality. The same scientist who described modern biology as collaborative also admitted: "Not every good scientist is a good collaborator." But, some of the firm's good scientists should be good collaborators, and program leaders or project managers must be good collaborators (and skilled at developing collaboration), because they are the linchpins in both internal and external efforts. One president of a biotechnology company stated the following:

> Project leaders, the people who manage target-directed drug discovery programs, must consider collaboration as a very important dimension of the program.

This attribute is equally important to the pharmaceutical partner. One vice president of R&D in a pharmaceutical company typically engaged in 20 or more strategic alliances emphasized the following:

> We hire people who are collaborative, and I spend a lot of personal time interviewing people as they come into the organization, to make sure they understand that they will be required to collaborate.
>
> I tell them that, for us to be successful, we have to bring a lot of disciplines from different areas and work together. We beat our competition by making sure these groups work together efficiently. We gain competitive advantage by efficient collaboration.

Ensuring that a formal recruiting system or general practices will support effective collaboration involves two important steps. First, the company's explicit or implicit recruitment screens should include the attribute of willingness to collaborate, ideally evidenced by *prior effective collaboration*. Interview guides should explicitly list "collaboration" as demonstrated by the candidate's track record, references, and unrehearsed conversation about prior experience. Those who will interview new scientists should agree on what they will take as indications of successful collaboration, before the actual meeting. Then, when they later discuss their impressions, there will be a basis for assigning judgment.

Second, preparing for effective collaboration does not end with recruitment but must address the way in which new hires are oriented to the company. Training (formal and informal) and mentoring must encourage "collaborativeness," and those responsible for overall orientation should be measured according to how well commitment to collaboration is institutionalized.

Performance Appraisal and Reward. Like recruitment systems, performance appraisal and reward systems must support collaboration. Systems that measure and assign rewards only (or primarily) for individual effort, or that use accomplishments and contributions of solo performers as comparison and example, will eventually stamp out whatever cooperative, collaborative spirit might have existed. As many managers discover, the behavior they reward is the behavior they get. If these systems and practices do not take account of collective efforts and reward collective performance, as appropriate, then collaboration will not be valued. What is necessary is a practice that appraises and rewards group and team skills, group and team level performance, and individuals' abilities to support and maintain collaboration.

Communication. The peer relationships that are supported by lateral structure are associated with open, candid, and informal communication. That type of communication will (by definition) provoke fruitful debate and intellectual challenge. Project information will be widely shared. Learning will take place, value will be added, and the collaboration will be effective.

The consequences of a vertical pattern of relating and communicating are far-reaching. We noted earlier that the structure of Lucida had a number of troubling signs of being vertical. In the course of the Lucida–Pharma Sciences alliance, we know that, at least initially, information was not shared (the chief scientist, O'Brien, did not give needed data to the project leader). The vice president of research, Santoro, also described alliance communication as inhibited:

> Ostensibly, there was a core team in-house being coordinated by the program manager. Because he was not supported, people . . . would show up at the biweekly meetings with Pharma Sciences, but they spoke independently, without vetting things with their group. Usually, our scientists ended up arguing with each other in front of the Pharma scientists.

Decision-Making. Decision-making practices must also be appropriate for collaboration. They should be straightforward and streamlined and should reflect trust in the intelligence of the people involved. Even in small organizations, it may be surprising to discover how many items or decisions that should not require formal decision processes are stuck in such systems. If decision-making is less than efficient within the organization, consider how inefficient it will be across the alliance.

Decision-making practices should be inclusive rather than exclusive, especially when broad-based commitment and action are meant to follow. Even when decisions are appropriately made by a single individual or small group, input and feedback from others are vital to gaining their support of the process and acceptance of the outcome. Organizations that have a more vertical than lateral structure are likely to have a more unilateral approach to decision-making (the implied hierarchy). In an alliance, their "side" may end up unintentionally making decisions alone, thus provoking fallout from the partner organization. Lateral structure and inclusive decision-making are prerequisites to alliance effectiveness.

Culture

We defined culture as shared values, beliefs, and norms and used the model of concentric circles to describe sector culture (biotechnology, pharmaceutical), organizational culture, and discipline culture. In the discussion that follows, we focus on organizational culture.

As might be expected, organizational culture has a profound impact on collaborative effectiveness. Our purpose, here, is not to provide advice on how to deal with aspects of culture that are inconsistent with effective collaboration. Rather, we highlight the links between culture and commitment, structure, and systems. We encourage readers to "seek professional help" for any problems that may emerge from a similar review of these issues in their own organization.

Commitment. Commitment to a specific collaboration is a necessary but insufficient condition for effective collaboration. People must be committed to the value of being collaborative. Thus, culture (as shared values and norms) must facilitate being flexible, tolerating and encouraging working together, taking risks, and handling failure well. The organization's culture must enable people to take intelligent actions that are different from what is customary, must allow for failure, and must support learning rather than punishing.

Structure. Lateral structure allows for (and supports) effective collaboration. That means, in effect, that organizational culture must foster beliefs in equality, rather than superiority and inferiority. There must also be beliefs in the distribution of knowledge and competence throughout the organization, rather than limited to positions of title and rank. By the same token, flexible and permeable borders between groups, disciplines, and departments facilitate effective collaboration. Norms in the organization must therefore discourage the demarcation of fiefdoms, kingdoms, territories, and all such indications of rigid demarcation that separates people.

Systems. If one of the shared values is "being collaborative," then recruiting norms will favor selection of people who are willing to collaborate (who share, in other words, that same value). The taken-for-granted routines of orientation and training will also foster this propensity.

Performance appraisal and reward systems as actually implemented (as opposed to what is written, if anything) are clearly reflective of a culture of collaboration or not. If being collaborative is valued, then performance measures will take this into account and the rewards will be consistent with that value.

If, instead, there is shared belief in the primacy of individual achievement, if working for one's own accomplishments is a widely held norm, then no amount of persuasive rhetoric about cooperation can hide the cultural reality.

Just as lateral structure depends on certain cultural attributes, so do open, candid, and informal communication systems. Belief in the efficacy of intellectual challenge is a cultural requirement for effective collaboration, as is the value of debate and wide information sharing. Culture is at the heart of structures that produce effective communication; it is also at the heart of structures that inhibit communication among the key players.

Finally, decision systems are also influenced by culture. Systems that are cumbersome, slow, and reflective of a lack of trust are in large measure the outcome of shared beliefs in superior–subordinate relationships. Systems that are exclusive rather than inclusive reflect norms that support hierarchy. Systems that are unilateral rather than joint or mutual reflect, again, a value in presumed superiority. Cumbersome, exclusive, and unilateral decision-making systems jeopardize collaboration.

CLOSING THOUGHTS

We cannot overemphasize how important preparation is, particularly for the biotechnology side. It is on the shoulders of biotechnology leaders that the

Covert leadership.

larger burden of responsibility for alliance effectiveness rests. Choosing the right people to fill critical alliance roles, along with having the right structure, systems, and culture, will make the day-to-day challenges of alliance leadership and management that much easier to handle. People in the biotechnology firm must assume responsibility for alliance effectiveness, regardless of how formal leadership roles and management functions are distributed. We argue that biotechnology leaders must sometimes practice *covert leadership*, working to enhance alliance effectiveness and to achieve alliance success, without benefit of glory, or even of recognition.

Of course, it also makes sense for the pharmaceutical side to be prepared, and the better their preparation, the better the alliance will work. (We discuss the latter subject in our Appendix to Chapter 6.) The harsh truth, however, is that the pharmaceutical side rarely needs the alliance in the same way.

Think differently . . .
        Lead differently . . .
                Make alliances work.

CHAPTER 6

INDIVIDUAL AND ORGANIZATIONAL DUE DILIGENCE

Getting the biotechnology "house" in order is groundwork that must be laid in preparing for collaboration generally. When a specific alliance is to be undertaken, biotechnology leaders must also conduct an assessment of the partner. The purpose of this assessment is twofold: to understand the partner better, and to be prepared for interacting in a way that ensures alliance effectiveness and, thus, alliance success.

Ensuring your own readiness to collaborate is a necessary but not sufficient condition of alliance readiness. Similar to the traditional due diligence process, assessment of the partner is intended to uncover information that is needed to move the alliance process forward. Due diligence, as we address it here, focuses not on scientific, technical, and legal matters but on alliance-relevant systems, practices, and people. This type of due diligence is conducted after the deal has been signed and the decision to collaborate has been made.

In this chapter we first discuss the fundamental issues of strategic fit and top management commitment. Then, we describe the issues and actions relevant to individual due diligence and to organizational due diligence. The Appendix describes due diligence from the perspective of the pharmaceutical partner.

Due diligence: systems, practices, people.

FUNDAMENTAL ISSUES

Strategic Fit

From what we know of the background to the Lucida case, assessing the strategic fit of the alliance took place predominantly within Pharma Sciences. The experience of their CEO, Dean, on the Board of Lucida exposed him to that company's strong patent position. He realized, also, the complementary expertise Lucida's scientists might bring to a Pharma Sciences' research program. At Lucida, the assumption seemed to be that, because this was Dean's "personal project," it was a good strategic fit for Pharma Sciences.

Even if we assume that a deal has been signed, it remains a responsibility of leaders on the biotechnology side to review the fit of the alliance with the strategic objectives of their partner and *to monitor that fit over time.* In a large pharmaceutical company, strategy will be formally articulated, codified, and revisited and revised on a regular basis. Part of the pharmaceutical company process will involve identifying strategic questions that must be answered. As our interviewees from large companies all noted, there are many such questions relevant to an alliance:

> . . . We ask: Can we sell the outcome of the alliance? Will managed care buy it?

. . . There are many business questions we ask about biotech: Will an alliance be more profitable for us in the short term? Is there clear business logic to the alliance? Are the numbers OK?

One senior pharmaceutical executive emphasized the following:

> The biotechnology partner has to get involved early on with what the expectations are with regard to the alliance. They need to have a good understanding of the criteria we use to put resources into an alliance—and to change the allocation up or down—and how we make decisions to go forward or not.

We describe the assessment of partner strategy (and "how we make decisions") in the section on organizational due diligence (under "strategy" and "systems," respectively). Here, we only want to reiterate the importance of considering and monitoring fit from the perspective of the partner. If the deal has been signed, then the partner's "deal signers" (at least) believe in strategic fit. But, determining what that means, how the alliance might add value to the partner, and how that value could change over time is a key responsibility of people in the biotechnology company. Executives in the large partner will be managing a sizable portfolio of outside collaborations, sometimes with several in the same therapeutic or technology area. Thus, both sides must understand the distinguishing attributes of a specific alliance and the strategic fit in terms of that portfolio.

Commitment

Of course, it is impossible to ask the partner if they are committed to the value of collaboration. (Who would answer: "No!"?) But, it is not impossible to find out from others who have collaborated, or know of collaborations with that firm, whether the value is really held. Some companies have a reputation for being difficult; in other words, for not valuing collaboration and what it entails. According to one of our interviewees from a biotechnology firm, their (difficult) partner's philosophy appeared to be the following:

> "I'm the best in the world. You need me. I'll give you the money, but I'm in it for everything and will screw you forever. The fact is that my mere presence makes it better for you!"

In contrast, a biotechnology scientist described their partner's implicit commitment to the value of collaboration as follows:

They behaved with us as if to say: "We're committed to being a partner with you. We'll stick it out, and we're prepared to give up assets for this partnership."

Finally, consider these "tests" of commitment, based on the wide experience of a biotechnology executive:

I define a "lousy" partner as one that looks at the contract every day and tells us we can or cannot do something. A good partner says about something that comes up: "This could be interesting; let's try it." A lousy partner just holds you to the letter of the contract, or they start worrying about their royalties before the science is even in order.

If the partner does not have a reputation for commitment as we have defined it (i.e., to the value of collaboration), then leaders on the biotechnology side are responsible for anticipating the consequences and preparing for them. Even if the partner is "lousy" in some ways, such as holding collaborators to the letter of the contract, leaders who prepare response scenarios will be more effective than those who are caught unaware.

DUE DILIGENCE

In this discussion, we are assuming that people in the biotechnology company have never worked with the large alliance partner and that there is much to be learned. Although the Commitment Champion in the biotechnology company will have some information about the partner when the deal is signed, there will be a gap between what is currently known and what should be known in order for the alliance to be effective. The largest amount of information resources—people, time, money—is needed at the start of the alliance, to accomplish individual and organizational due diligence.

Individual Due Diligence

The goal of individual due diligence is to learn about the partner's key alliance people. There are usually several individuals who should be identified early because their efforts, actions, and commitment will have a sizable impact on the effectiveness and success of the alliance. Certainly, their project leader (sometimes called team leader, alliance manager, project manager, etc.) is a key person, because this individual will be responsible for leading the pharmaceutical company's day-to-day alliance activities. At the executive level, knowing who is serving in the alliance champion role is critical. One of our interviewees suggested the following:

> It should be clear from the authority lines in the company who are the principals with whom we should do business. These are the people who are empowered to say "yes." We have to understand who can make the decisions and really assign the FTEs to the alliance effort.

Often, the person who fills this alliance championship role will be one of the therapeutic area vice presidents. The individual may be deeply invested in the alliance and willing to work hard and "cheerlead" loudly to facilitate effectiveness and success. Ideally, said one biotechnology scientist:

> There has to be a belief that this alliance is a really important thing, and we [i.e., the large partner] are doing it because there is the potential for a leap forward. We're not doing it because it's additive. [He continued: Anyone who thinks "additive" in the large company should be shot and should not be involved in the collaboration in the first place!)

Individual due diligence.

Unfortunately, it is also possible to find that a nominal "champion" has been assigned the task and does not truly believe in the specific alliance or in the general value of partnering. To determine the champion's likely level of investment, biotechnology leaders should find out about the history of the deal. Whose idea was it? Who was involved early on? Have any of the initiators changed jobs, been promoted, left?

In addition to the people with formal titles or nominal responsibilities, there will be another handful of people who matter. These are the individuals whose input to alliance decisions is sought and/or seriously considered. Biotechnology leaders should identify those people whose support for the alliance can make a difference in how the alliance progresses over time.

On a day-to-day basis, the team members on the partner side are critical to the success of the work. A biotechnology scientist said that once individuals have been identified,

> . . . we do a "scientific survey" of our collaborators. We find out what they have published. If the culture of the company does not support publication, we have to go to secondary sources, such as SCIFI. This tells us everything they have done, from the thesis forward. You'll still have an incomplete picture, even with publications, so we use the *old boy network* and ask our contacts: "What do you know about this person?" We want to know the person's track record, from publications as well as scientific reputation.

Organizational Due Diligence

Everything that can be learned about the partner company is useful. Some information, such as reputation, can only be discovered by listening carefully to those who have prior experience collaborating with the company (or know people who have that experience). Other information can be learned during the initial meetings undertaken to reach a verbal understanding, as well as during formal contract negotiations. No opportunity to observe and reflect on the partner's attributes is too small to forego or to ignore. As one of our biotechnology interviewees said:

> In our meetings with the partner, the first thing we do is try to discover where the power is. "Knowledge is power." It's pretty simple to take that dictum and look for the most knowledgeable people in the room on a subject, and you'll know where some of the power is in the organization.

Organizational due diligence.

In the remaining discussion, we address what we believe are the key elements of organizational due diligence: strategy, structure (specifically the matrix, which is common in large companies), and systems (especially portfolio decision systems).

Partner Strategy. Facts about the partner's strategy can be gained from numerous sources. As soon as the deal has been signed, a responsible individual from the biotechnology company should acquire the last two or three annual reports of the company and study them. Typically, the task of learning from the annual reports would be done at quite senior levels, perhaps by whoever is serving in the Commitment Champion role. Important information includes the drugs being sold or developed in the therapeutic area relevant to the alliance and what the company is saying about them. (Another source is the company's website.) Recent articles about the company can be found, using a major search engine service, as well as reports on the company written by Wall Street investment firms.

These sources provide the basis for understanding the large company's portfolio of marketed products and of drug candidates in later stages of development (Phases II, III, and in registration). They also allow the biotechnology partner to assess how the alliance invention is intended to fit into the overall portfolio. Will the invention allow the partner to enter a brand new therapeutic area ("first in class" therapy)? Will it provide a useful treatment in an area where the partner already has products? Will it allow the partner to "catch up" to the competition—or leapfrog them—in terms of products or technologies?

Answers to these questions can only come from a thorough understanding of the therapeutic or technology area, knowing what is currently being used and what is worked on in the industry. Data on a therapeutic area, products, and comparative strengths and weaknesses can be found in review articles from the literature or from series such as *Medicinal Research Reviews* and *Annual Reports in Medicinal Chemistry*. Over the course of the collaboration, both partners must be able to discuss potential competitor products or technologies that are still under development, and when they are likely to reach the market. Informed discussions will help to maintain a realistic perspective of the value of the partnered invention.

For the sake of illustration only, we will take Merck and show some examples of what publicly available data can reveal.

The company's 1998 annual report contains this statement of strategic objectives: "The goal of Merck research is to introduce innovative, unique-in-class medicines *that can be taken orally, once a day and are well tolerated*" (emphasis added). The end of the sentence might imply that parenterals, injectables, oral drugs that have to be taken several times a day, drugs that have obvious side effects (despite their efficacy for the stated indication), and so on are not likely to be considered a best fit with Merck's strategy.

The annual report provides a table of Merck drugs, from those in Phase IIb studies to those on the market. The financial section includes sales data from 1996 through 1998 for eight therapeutic areas. Revenue data, plus the lists of products on the market, plus statements in the text (e.g., "*Vasotec* and *Prinivil* account for 30% of the world market for ACE inhibitors") indicate where senior managers might be most interested in terms of alliances. For example, combined sales of Zocor and Mevacor (for hyperlipidaemia) accounted for almost 40% of Merck's total pharmaceutical sales of $12.8 billion. Also of importance are patent expiry dates (in the financial section), because the company will be pressured to have replacement drugs that can sustain revenue growth and profitability.

Some of Merck's current licensing agreements are listed, including the name of the organization and the therapeutic area under contract. Such information allows senior management in the biotechnology company to talk to their counterparts who have had experience with this company. The *Wall Street Journal* reports on companies' quarterly earnings and their pipelines (these articles may contain links to more in-depth reports). Consulting companies also publish news briefs that may be useful. For instance, Scott–Levin consults to pharmaceutical companies and publishes short articles on its website. In one of these, Merck was named as the company with the best training program for sales reps, suggesting that a Merck salesforce could be very effective in marketing a drug. Fund managers who invest in pharmaceutical companies will publish their assessments. Because there is so much money at risk, it is in fund managers' best interest to investigate their stock picks rigorously.

***Partner Structure (The* Matrix).** Earlier, we described two general characteristics of organizing work that support scientific and alliance effectiveness: lateral direction of relating and communicating, and flexible and permeable demarcation of groups. These characteristics of how alliance tasks are organized (i.e., alliance structure) are vital. Our purpose is not to give advice on how to modify the partner structure but rather to explain the probable partner structure, the *matrix* organization. Many large

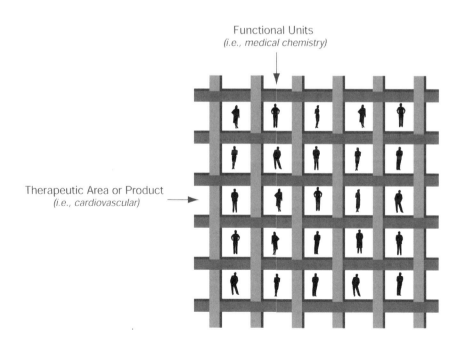

Functional Units
(i.e., medical chemistry)

Therapeutic Area or Product
(i.e., cardiovascular)

companies, and most large pharmaceutical companies, have a formally articulated matrix organization. From the biotechnology perspective, we believe it is important to understand what such a structure is and what it might "feel like" during the alliance.

A matrix is defined as a network of intersections of vertical and horizontal lines, similar to the warp and woof in a piece of cloth. In a company, the analogy to the vertical "line" is usually the functional or discipline units, such as "medicinal chemistry" or "pharmacokinetics," and so on. The analogy to the horizontal line is the therapeutic area or product or technology unit, such as "cardiovascular" or "genomics."

Project teams in large pharmaceutical and biotechnology companies are customarily organized within the matrix. Team members come from, and report on technical issues to, the functional and discipline units. They also report to the project leader or therapeutic area head on project issues. Each team member, in effect, has two bosses: the discipline boss, and the project leader as representative of the therapeutic area boss. Within the team, however, members will relate to each other and communicate with each other as peers. Authority will shift as problem expertise shifts. No one discipline can dominate; all are needed to complete the task.

In the large partner, a person with, say, expertise in bioinformatics might be assigned to a "cancer gene therapy" team. That person, then, will work at the intersection of the bio-informatics group (vertical) and the gene therapy project within the cancer therapeutic area (horizontal). The bioinformatics team member will be responsible to the project leader for achieving, equally with other members, the project charter (e.g., the goal of moving an alliance invention towards practical gene therapy). At the same time, the bioinformatics member will be responsible to the section head for the caliber of his/her technical contribution.

In large companies, the collection of projects may be organized as a matrix as well. That is, there will be a project management function on the same level as the other functions and disciplines. In our example, the project manager for the cancer gene therapy team will then be responsible to the head of project management (rather than the head of cancer therapeutics) for completion of the project on time and within budget. The project manager will be responsible to the functional heads for effective deployment of their resources, particularly the team members. The head of project management will work with therapeutic area heads, to ensure that all projects (including cancer gene therapy) meet corporate strategic objectives.

The matrix structure is intended to ensure equal contributions of technical and project expertise as needed. Therapeutic discovery and development involve numerous complex tasks and many disciplines and are characterized by high scientific and technical uncertainty (and risk). These are the conditions under which project management and matrix structure are more effective than any other organizing framework.

But, as this illustration should underscore, the matrix structure is very complicated. In reality, rather than representing functions and products equally, the structure may be lopsided. Either technical and discipline managers dominate decisions and resource allocation, or product managers dominate, and who dominates may change over time. Studies of matrix structure suggest that it works best in organizations that are neither so small that such an arrangement would overcomplicate the work nor so large that coordination in both directions at once is overwhelming. In addition, people need excellent interpersonal skills and conflict-resolution expertise. The biotechnology company would like to know whether their partner supports the development of these skills and abilities through the promotion of informal and/or formal training.

If an alliance team includes members from a large company who are "embedded" in a matrix, that complexity must be taken into account. In the biotechnology company, the alliance may be the primary or only project at one time. Team members may be wholly dedicated to that project, or nearly 100% assigned to that project. The project leader on the biotechnology side may report directly to the Vice President of R&D, or the Chief Scientific Officer. In the pharmaceutical company, the alliance may be one of many in a portfolio of external collaborations (that is, not even including internal efforts). Team members may be assigned for fractions of their time to several projects. And, the project leader may report to a head of project management overseeing tens (if not hundreds) of competing projects.

These facts argue for formal project management training of those in the biotechnology company who will work on that complicated project interface, particularly the Process Champion. As we have emphasized throughout this book, most of the burden of preparing for the collaboration falls on the shoulders of people on the biotechnology side.

Partner Systems. We addressed recruitment, performance appraisal and reward, communication, and decision systems within the biotechnology company. Here, we focus on portfolio decision systems that are, like the matrix structure, customary in large companies. With regard to the other systems, we will only summarize a few important issues.

Recruitment. The biotechnology partner cannot influence who is recruited from the pharmaceutical company. But, people can inquire about the orientation and training of new staff, particularly with regard to collaboration. The objective is to be prepared to accommodate what cannot be changed.

Performance Appraisal, Reward. The issues are the same—ask about what is measured (individual publications and patent applications, or contribution to collective problem-solving) and be prepared. If partner scientists are rewarded primarily for individual accomplishments, then great care has to be taken to ensure that team members really work together. In the words of one interviewee, if the company doesn't really value and reward working together, you'll have "everyone doing their own thing":

> There is a lot more pressure to make sure everyone knows "I'm performing" than to collaborate. "I have to have credit!" There is no cooperation because of the pressure to shine as an individual. Most people here would still follow a tangent that resulted in a paper than do work that would move the project forward. To be successful here, you have to demonstrate your success with papers published (pharmaceutical scientist).

Communication and Decision-Making. What the biotechnology partner must typically accommodate is the lengthier "communication travel time" that may characterize communication within their partner company. One of our interviewees from a large pharmaceutical company cautioned:

> The expectation from the biotechnology side is colored because they think: "I just have to go up and ask Fred in the office and he says OK." Whereas, in the collaboration they find they are calling, and calling, and faxing, and e-mailing this big company, and they do not get a quick response each time.

> The little guy has very different expectations about communication and the speed of decisions and the flow of money . . . The project interface [between biotechnology and pharmaceutical organizations] is complicated.

Of course, the longer the communication travel time, the longer it will take to make many different kinds of decisions. Even decisions that might seem straightforward may need to be passed along a communication, so that everyone is kept in the loop.

Portfolio Decision Systems. The objectives of a portfolio are to (1) meet agreed-upon goals and (2) balance risk of failure. Senior executives of large companies regard drug discovery efforts and the collection of projects (in-house as well as external) as a portfolio of assets that must meet strategic goals. Novel efforts are highly risky; they have to be balanced, therefore, with sufficient projects that have higher probability of success.

The difficulty with any type of portfolio (stocks, bonds, money, biomedical discovery efforts) is the same: lack of certainty. The more speculative the effort or project, the less predictable are the technical and marketing feasibility, costs, return, time to market, and competitive status. Judgments on costs and revenues are based on an assumed monetary outcome multiplied by the probability of that outcome. Judgments on technical and marketing feasibility and competitive status are also based on assumptions.

What people in the biotechnology company must realize is that the alliance is just "some number in the queue" of a large portfolio of discovery efforts and projects that are all competing for a finite amount of corporate resources. In the partner firm, managers are continually (1) looking for ways to ensure that projects are aligned with strategic objectives, (2) taking advantage of tradeoffs between projects (e.g., slowing down or stopping some in order to "fast track" others), and (3) balancing the portfolio along the dimensions of benefits, risks, and costs.[1]

The Lucida–Pharma Sciences case provides an illustration of portfolio decision systems at work, to the detriment of Lucida. One of the Lucida senior scientists said:

> . . . [early studies on the compound] were not consistently reproducible. So, Pharma asked for some repeat studies in outside labs. When those results came in, they were ambiguous. It took weeks before we could understand what was going on.

[1]Philips, L. D. *Approaches to Prioritizing Projects and Creating Portfolios*, London: Facilitations Ltd., 1994.

People in Lucida were apparently not aware that, within Pharma Sciences, senior managers were balancing the portfolio and "taking advantage of tradeoffs between projects." Delays in the Lucida alliance project prompted managers to "shift their R&D allocation to compounds with greater likelihood of near-term success" (published article on Pharma Sciences' decision to terminate the collaboration).

Most large partners will use a portfolio approach to R&D efforts and will use a sophisticated and quantitative methodology for portfolio decisions. If their external collaborations are numerous, they may have a "subportfolio" devoted to these projects. That subportfolio may be subject to tradeoff with in-house efforts, further complicating the situation for the biotechnology partner. Given such a system, it is even more important that people in the biotechnology company be prepared to monitor the strategic fit of their invention, to understand the distinguishing attributes of that invention, and to "get involved early on with what the expectations [of the partner] are with regard to the alliance." From what we know of the Lucida–Pharma Sciences alliance, that never occurred.

CLOSING THOUGHTS

We believe that organizational due diligence does not end here but also requires an assessment of the partner's organizational culture. Assessing partner culture begins with the first contact. Whenever we have addressed company "reputation" (as in *reputation as a collaborator*), we have implicitly addressed culture. It is the responsibility of everyone in the biotechnology company to pay attention to signs of shared values, beliefs, and norms. As one of our interviewees noted:

> Even in the earliest scientific meetings between the two teams, we watch for their philosophy or culture about strategy, and where they see their own competence. We watch the interaction among their people, to try to understand the "style issues."

> Some companies send a lot of people to team meetings, because there is a lack of trust. In other companies, we've noticed that the chemists won't say anything in front of their boss, and that tells us something. In some companies there is a lot of internal squabbling, and we've learned to steer clear of that. And, we've had experience with teams consisting of "delicate flowers," or large egos that had to be handled with care.

Appreciating at least some of the aspects of partner culture allows people in the biotechnology company to consider a range of responses and to adjust to (or accommodate) partner behaviors. Perhaps most importantly, appreciating the differences in company cultures can reduce the misunderstandings and miscommunications that might otherwise impede effective collaboration.

> Think differently . . .
>> Lead differently . . .
>>> Make alliances work.

APPENDIX

In a departure from our focus on the biotechnology partner, here we address due diligence from the perspective of the pharmaceutical company. In this discussion, we switch hats, taking the role of the large partner. Moreover, we move back in time, to before a deal is signed. Our purpose is to illustrate some of what precedes the signing of the deal, particularly the critical process of reviewing a biotechnology firm's scientific and technical position and data. Our shift to the pharmaceutical partner is important. The reality is that people in the large company complete the traditional due diligence, and senior pharmaceutical executives say "yea" or "nay" to the alliance.

Preparation Within the Large Partner

There are preparatory steps that should be undertaken by the large company to maximize the chance of alliance success, to make the collaboration as efficient and effective as possible, and to minimize the interpersonal problems that inevitably will arise. In a role similar to the Process Champion in the biotechnology company, the large partner must also have a *champion* for the alliance. When the likely invention is identified, an individual should be selected, based on both technical competence and leadership skills. He or she should be clearly identified and announced to the organization. Then, the champion must set out immediately to accomplish three key tasks: (1) ensure that there are advocates for the alliance within in-house discovery, (2) develop a process for determining the value-added of the alliance, and (3) design a framework and process for completing systematic due diligence on scientific data presented by the biotechnology company. If more than one division in the large company will be affected by the alliance, the champion must ensure that cross-divisional consensus will be achieved at every critical node.

In-House Advocacy. The champion should begin to consult with all appropriate members of the discovery department, at all levels of management and bench scientists, in order to avoid the "NIH" (Not Invented Here) syndrome. In-house discovery scientists will naturally be more supportive of their own ideas, projects, and drug candidates. Any new technology or drug candidate brought in from the outside will likely be viewed with at least skepticism, or perhaps even open hostility. The champion should identify who in discovery needs to be approached, and then explain the reasons for the alliance and the advantages it brings to the company and to that particular discovery scientist. In order to make allies of those scientists, the alliance champion must understand that discovery scientists will be interested in "what's in it for me?"

The champion should endeavor to have a discovery scientist become an advocate for the partnered invention. The rewards of advocacy include sharing in the recognition and "perks" (such as lecturing internationally, attending symposia, broadening his/her reputation) that accrue to successful collaborations. Moreover, the unique, scientific properties of the proposed device, technology, or medicine hold promise of improving therapy in an important way. These advantages (what one of our biotechnology interviewees rightly termed *the vision of what the science can provide*) cannot help but raise the level of excitement about the partnership.

Of course, to present the exciting features of the partnered invention, the large partner champion must have met with key scientists at the small biotechnology company. An important purpose of such exploratory meetings is to marshal the information needed to make his or her arguments "back home."

Alliance Value-Added. Because the large partner is managing a sizable portfolio of outside collaborations, several of which may be in the same therapeutic area as the invention at hand, the champion must understand the distinguishing attributes of the alliance outcome. Like the biotechnology champions, this individual must know the likely strategic fit of the invention within the company's portfolio, in order to "sell" it to the organization. In addition, the champion must develop a process for determining the value-added of the alliance *over time*. New data will influence the perception of value; thus, value is never set "once and for all" at the start of a collaboration.

The champion should begin to identify qualified people who will help with the crucial task of estimating the commercial value (net present value, or NPV) of the technology or drug candidate to be brought in by the alliance. Beyond the calculated NPV, of course, it is expected that the value of the invention will also derive from less quantifiable factors. These include:

- The invention's ability to supplement or complement a product already sold by the partner.
- The invention's ability to give the partner a competitive advantage in a therapeutic area in which they already have products.
- The invention's ability to match or preempt a competitor in such an area.

Sometimes, NPV may be a relatively modest amount, and these other considerations can be overriding factors in determining the value-added of the invention. The champion, in consultation with experts in the biotechnology company, is responsible for identifying where exactly the value of the outcome will reside and for monitoring changes in perceived value over time.

Scientific Due Diligence. Related to the process of determining and monitoring alliance value-added is this third task of the in-house champion: designing a framework and process for completing systematic due diligence on scientific data presented by the biotechnology company. Detailed review of new data (*scientific due diligence*) is a major part of what moves any alliance forward. Thus, the champion must identify team members who are appropriately qualified to participate in scientific due diligence, including scientific experts, marketing personnel, and legal and patent attorneys.

"Due diligence" is the term given to the process by which information and data, previously merely outlined by the company owning the invention, will be reviewed and verified in great detail. The inventor company must produce all the records by which the alliance partner, under a secrecy agreement, can review the raw data pertaining to the invention. Mechanical

Scientific due diligence.

innovations may be comparatively easy to evaluate—simply answer the question, Do they work? But, novel medicinal agents can undergo very complex technical data reviews that will grow even more lengthy when the drug candidate is in later stages of development.

Some examples of questions that might be asked could include:

- Is the chemical structure and purity of the compound firmly established?
- Were appropriate doses and species used in animal models of efficacy?
- Were blood levels of a drug measured?
- Are the animal models well established and accepted in the field, or are they unprecedented?
- Was there sufficient power built into experiments to allow for a significance level of $p = 0.5$ or less?
- In terms of toxicology data, were appropriate species used? adequate doses? postmortem tissues examined?
- In terms of "good laboratory practices," were experiments conducted under GLP conditions?
- If manufacturing was involved, what is the largest scale synthesis conducted to date? Can the current process be scaled up?
- Is a suitable clinical formulation in hand? Will human pharmacokinetic [PK] studies need to be done?

- Finally, even at a very early stage of the alliance, clinical trial design should be considered. Who are the target patients? Will they be difficult to recruit? What dose levels should be explored? Do the necessary clinical endpoints have precedents? Are there sufficient clinical supplies and placebos available for blinded studies?

The Lucida–Pharma Sciences Situation

Although we have only the perspective of Lucida scientists and executives, we can infer that the large partner's responsibilities were not as thoroughly accomplished as they might have been. Let us start with in-house advocacy.

A Lucida scientist felt "a certain amount of resentment on Pharma Sciences' part" when the collaboration began. He observed, early in the process, a notable lack of enthusiasm (let alone advocacy) for the project:

> I think there was a certain amount of resentment because we asked for too much in their eyes. There was not, per se, the *NIH* syndrome. But, something was not right!

On the other hand, one of his colleagues said:

> This was a top management deal. When the program was first announced, there was skepticism among the team at Pharma Sciences, some *NIH*, some discussion of "why are we doing this?" Because it was their chief executive's personal project, they felt they had to make *some* effort to see if the science would work.

If the Pharma Sciences' alliance team were not convinced of the *vision of what* [Lucida's] *science could provide*, then perceiving added value from the collaboration would be difficult, if not impossible. Compounding these problems (lack of advocacy, unclear added value) was the lack of systematic and timely due diligence—a failure shared by both companies. To Lucida team members:

> . . . if you talked with Pharma scientists, they would say the deal with Lucida was forced on them by senior management. They had questions that were not adequately answered, and they did not have time to ask questions that should have been asked.

Lucida scientists acknowledged that there was a "data dump" in the early meetings. And, partway through the attempt at due diligence, Pharma Sciences made a decision that was problematic:

> Pharma Sciences brought in a new project manager, and I think there was a certain amount of resentment towards that new person on their side as well. Altogether, there were two new people in the alliance, two new project managers, and communications were not terribly effective.

Because of resentment and ineffective communications, what should have been a collaborative endeavor became marked instead by "real bitterness between the two groups. . . .

> Pharma Sciences had paid a lot of money upfront, and their scientists kept saying: "What have we got ourselves into?!"
>
> I blame them for not doing due diligence, and I blame us for not reviewing our protocols well enough to spot the likelihood they we might get equivocal data.
>
> At the end of the day, I'm not surprised their senior executive actually said, "if there were a lemon law for biotechnology, this alliance would be eligible!"

CHAPTER 7

THE FIRST MEETINGS

In many ways, the culmination of efforts to prepare for collaborating is the first formal meeting. At this event, people from both partners will interact face-to-face, many for the first time. (Ongoing informal meetings are also important in the "getting to know you" process that must occur as the alliance moves forward.) At the first formal meeting, "live" impressions are important, if only because so much future interaction will be over a distance, via e-mail, LAN, telephone, and so on. Getting to know the "real" partner, and being known in the same way, must be choreographed and managed effectively. The first formal meeting represents a critical juncture in the alliance. If the meeting goes well, the collaboration is off to a good start.

For a number of players, this first formal meeting is not the first time they will have met face-to-face. Earlier in the due diligence process (scientific, technical, individual, and organizational), some of the leaders and scientists on both sides will have had opportunities to interact. During these meetings, information will have been collected, disseminated, and shared (in other words, uncertainty will have been reduced). These meetings also help both sides deal with the inevitable equivocality that surrounds the initiation of an alliance between partners who have not worked together before.

Our purpose in this chapter is to address the informal ongoing meetings and the first formal meeting, the kick-off event. Both kinds of meetings present opportunities to make the collaboration more effective. Alliance

The first meetings.

team members can begin to learn how they will work together. And, people from the biotechnology firm can begin to create relationships that will endure over the course of the alliance.

INFORMAL MEETINGS

Informal, face-to-face meetings are needed during the due diligence period and beyond. In addition to the written information that will be forwarded, discussed, and disseminated within and between firms, other information and understanding are required that cannot be derived from written sources. Building good relationships is so important to alliance effectiveness and alliance success that explicit steps like the following should be taken:

- When the small team of key people from both partners is identified, begin a series of informal meetings, over dinner or lunch, at alternating sites convenient for each company.
- Plan informal meetings up front, and clear everyone's calendar for those dates. Strive not to move these established dates, to show to everyone that these meetings and the alliance are receiving highest priority.
- Encourage general conversation that facilitates "getting to know you" between the key players. Rather than jump into business topics, allow time at informal meetings for free-ranging discussions of the background, training, and nonwork interests of each of the team members.

- Over time, build relationships by sharing information. Be open with each other about, for instance, who are the decision-makers in the respective organizations? What are the respective decision-making processes in the company? Will this collaboration face resistance from senior managers in either company?

- Begin to discuss how this group of collaborators will work together. Develop team rules, such as meeting frequency, dealing with lateness, accommodating absence, reaching agreement, and so on.

- Prevent inappropriate behaviors on the part of any of the advocates involved in the negotiations. It is not unusual for the inventor of the technology to be overly one-sided in support of the invention. Other team members must, in a sensitive manner, bring such a one-sided advocate to a reasonable view. The inventor must accept the fact that his or her "baby" (i.e., invention) will be in the hands of a team. Continually emphasize that the *data*, past and future, will drive the decisions by consensus at every step, and that this process is the most reasonable for both partner companies. Remember in the Lucida case that the lead bench scientist at Lucida, Will O'Brien, was a difficult person who was not a team player and never changed his opinion. Such a person, so valuable to the alliance, should have been confronted early on. Discussions of his vital scientific role and his need to work in a team environment might have prevented some of the subsequent behavior that contributed to the failure of the alliance.

- Make certain that all team members understand the value of working together toward the common goal—an effective and successful collaboration leading to a successful product. If any lukewarm or negative attitudes about the collaboration are detected among team members from either company, weed out these "naysayers" as quickly as possible. Opposing viewpoints are to be heard and considered, but the constantly negative person will be harmful to the process.

THE FIRST FORMAL MEETING

In the distributed environment of today's alliances, most of the interactions of alliance team members will be long-distance. Collaborators will be joined in a *virtual team* through e-mail, teleconferencing, telephone, and so on. Thus, as one executive from a biotechnology company emphasized:

There has to be enough face time with your partner up front, that you can keep the virtual team going. One of our scientists is a whiz at spotting body language and, once he knows the people, he is able to interpret the voice tones over the telephone. You have to be able to figure out each individual that way and understand what each wants out of the effort and, over the course of time, what are their concerns and their upsides. . . .

The single most important thing—no big surprise, of course—that makes these alliances work is the human chemistry.

Face-to-face interactions are the most effective way to maximize productive human chemistry. Such encounters also provide important opportunities to reduce equivocality (what our interviewee described as "figuring out" individuals and understanding their concerns).

The first formal meeting is not only a continuation of the prior "arm's length" efforts to reduce uncertainty but also, and most importantly, the beginning of the concerted effort to reduce equivocality. Individual and organizational due diligence efforts are primarily undertaken to reduce levels of uncertainty. For example, investigating a pharmaceutical partner's strategy clarifies (i.e., reduces uncertainty around) the likely fit of the biotechnology invention in the partner's portfolio of research activities. Inquiring about orientation and training of partner scientists clarifies (reduces uncertainty) their likely motives and incentives for collaborating. Asking how the partner is organized reduces uncertainty around the likely speed of communication and decision-making during the alliance.

Face-to-face meetings are the primary means by which people can begin to reduce levels of equivocality. Equivocality (ambiguity) exists when issues are unclear, when it is difficult even to articulate the right questions, when there is confusion, and when different people interpret the same information in different ways. Whereas uncertainty is reduced through information processing activities, equivocality is reduced through transactive (two-way) communication that creates shared meaning and understanding.

Simply disseminating information does not reduce ambiguity. Providing large amounts of scientific and technological data will not, by itself, ensure that results are interpreted in the same way. Remember how one of the Lucida scientists described the initial meetings between Lucida and Pharma team members:

When I look back to the initial meetings between the two groups of scientists, I realize that Lucida came in and did a "data dump" and *then* Pharma scientists began to work out just what was involved.

Lucida scientists provided quantities of data, assuming that Pharma scientists needed information. However, the Pharma scientists were confused about "just what was involved." In the context of uncertainty and equivocality, Lucida scientists may have reduced uncertainty, but they did not reduce equivocality. A lot of time was spent trying "to understand what was going on" in the Lucida–Pharma Sciences experiments, because of well-intended but mistaken efforts to reduce uncertainty (to get the facts) rather than equivocality (to reach agreement).

Questions remaining after internal preparation and organizational due diligence have been completed cannot be answered by collecting and processing information. Answers to questions about how team members should work together do not exist "out there." Asking what a partner means by "the value of collaboration" is guaranteed to provide incomplete information at best and misleading information at worst. Rather, agreement must emerge over time, through conversations and discussions and as people work together to reduce the ambiguity of roles, norms, and expectations.

We want to emphasize that equivocality is not reduced spontaneously with the passage of time. In fact, without explicit attention and effort to reduce it, the level of ambiguity may increase and hamper (if not thwart) the collaboration. Equivocality is an attribute of collaboration that, like the experiments themselves, must be intelligently managed. However, the time it takes to reduce equivocality can be compressed, to the benefit of the alliance project. Our objective in the remainder of this chapter is to provide guidelines for developing, in an effective and timely manner, the common understanding and interpretation of roles, norms, expectations, and work processes that will produce the right "human chemistry." It is at this first formal meeting that the equivocality about working together as alliance partners must be addressed.

General Guidelines

Before discussing specific issues relevant to the first formal meeting, we want to provide a few general guidelines for this and future meetings, virtual and actual:

- Consider carefully who should attend, but be generous for the first meeting in including all those who need to know about the alliance goals, activities, and so on. If you wish people to commit their time and energies to some aspect of the project, then you should include

Planning the First Formal Meeting
1. Be generous in inviting; include those who need to know about alliance, goals, activities, etc.
2. Choose a date convenient to as many attendees as possible.
3. Send out an agenda to all attendees ahead of time.
4. Specify in the agenda that time will be devoted to interacting and working in small groups.

them at the first meeting. The collaboration, after all, is just one activity competing with many other priorities for the attention and efforts of team members. Later meetings should include a smaller number of attendees, for the sake of efficiency.

- Choose a date that is convenient to as many attendees as possible, without delaying the meeting.
- Send out an agenda to all attendees ahead of time. Identify all speakers, along with their topics and time allotment (making sure each speaker is allotted a reasonable amount of time for presentation and discussion). Of course, the agenda should also identify who is responsible for organizing, leading, and coordinating the meeting.
- Make it clear in the agenda that time will be devoted to interacting and working in small groups.
- Identify the person responsible for action minutes to be written, reviewed (by whom?) and distributed (to whom?).

Ideally, both partners should share in the preparation and meeting tasks. However, people on the biotechnology side must ultimately be responsible for relationship management, and this first meeting will set the tone (so to speak) for future interaction.

One major goal of the first formal meeting should be to establish the climate of frankness and honesty that should characterize the collaboration. If one partner believes there have been attempts to hide problems or to use deceptive, devious methods, the collaboration will be off on the wrong "foot." In the case of the failed Lucida–Pharma Sciences alliance, there were technical and interpersonal surprises that undermined the collaboration. As recounted by one of the Lucida scientists, there was often an atmosphere of "Oops! I never heard that before!" Even if these surprises are unintended, they may damage the relationship.

In this initial meeting (and in succeeding meetings), speakers should make it clear whose views and ideas they are presenting. It is important to know whether the position/opinion represents that of senior management (implying previous consultation with them), or whether the position is that of a team or of the speaker alone.

The tone and feel of the first formal meeting are important. First, as the saying goes: "You have only one chance to make a first impression." Of course the biotechnology side wants to make a good impression, and the most positive adjectives describing a good impression may be *competent*, *cooperative*, and *congenial*. Contrast these adjectives with how Santoro described the early Lucida–Pharma Sciences meetings:

> . . . [People] would show up at the biweekly meetings with Pharma Sciences, but they spoke independently . . . [and] usually our scientists ended up arguing with each other in front of the Pharma scientists.

Second, behaviors and attitudes observed in the first meeting will be those others come to expect, and these expectations are a powerful determinant of behavior and action. One of our interviewees recounted the first meeting with a large company in this way:

> We thought, from all our prior preparation, that there would be five or six key people in that meeting, but there were 35, and they sat passively while

A Good First Impression

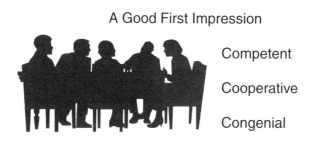

Competent

Cooperative

Congenial

we spoke. We knew immediately that this collaboration was jeopardized (biotechnology scientist).

Third, if successful, this first meeting can help establish the common ground—interpretations, meanings, and norms around work roles and communication—that will create the right human chemistry and support the virtual team over time:

> You want quiet time with individuals, so you can talk about science and the business types can talk about business. Some of our best conversations at the first meeting have occurred during the unscheduled interim before lunch or dinner (biotechnology project manager).

Below, we address the more specific issues relevant to the first formal meeting.

Purpose and Objectives

The overarching goals of this formal meeting should be to (1) provide information that people on both sides will need, such as experimental methods and data (thus reducing uncertainty), and (2) begin to create common ground amid the confusion that surrounds any startup project (thus reducing equivocality).

With regard to the latter goal, we appreciate that reducing equivocality takes time. People have to get to know each other, share their thoughts and ideas, and develop norms for working collaboratively. At the first meeting, however, setting specific behavioral objectives for how people will relate to each other helps build desired collaborative norms. We believe that a minimum set of objectives should include the following (we also provide examples of what should not happen):

- Candor (no hidden agendas, no surprises, no hedging . . .)
- Respect (no personal criticism, no sniping, no sarcasm . . .)
- Active listening (no interrupting, no responding on a completely different topic, no ignoring a speaker . . .)
- Open challenge for clarification (no agreeing for the sake of "peace," no avoidance of appropriate questions, no side bar discussions of issues . . .)
- Collegiality (no isolation of individuals, no cliques, no "in" versus "out" groups . . .).

This meeting provides an important opportunity to model desired behaviors and to begin building good relationships among team members. Senior managers should openly offer statements of support for the alliance, its goals, and its timeline. They might also make a presentation about formal alliance roles, responsibilities, and reporting relationships, while acknowledging that these may be adjusted over time. Such an introduction will help to answer some of the questions that scientists will have about working together and about who needs to know what (in other words, this will begin to reduce some of the team's uncertainty). Then, people must be encouraged to work together and get to know each other in small groups.

Of course, the first formal meeting is also about the work of the alliance; thus, uncertainty and ambiguity in the scientific/work domain must be addressed. Each side should make presentations and allow ample time for questions about what has already been accomplished, what kinds of work each side may need to do, and so forth. These activities lay the technical groundwork for the research plan:

> In a good collaboration, you set goals and you define what it will take to be successful, so you know when you're crossing the goal lines. A good agreement is one in which the research plan has been worked out in detail upfront (biotechnology project leader).

There should be an open exchange of information about scientific, technical, and even marketing issues related to what might be a final product, process, or other invention. Key issues should be raised, and questions posed by meeting participants should be answered—if not immediately, then soon after the meeting. Managers and scientists from the biotechnology company should present data in a concise, graphically pleasing format, and they should be prepared to be challenged. This is a normal part of scientific interchange and should not be taken as criticism of the ability of the presenter. Very detailed discussions (e.g., toxicology results, chemical processes) can be deferred to subcommittee meetings. But, all potentially problematic issues must be raised at the first meeting. There should be no big surprises for subsequent meetings (e.g., "we have not done any of those experiments yet").

Attendees

The people who will commit their time and energies to some aspect of the project should be included at the first formal meeting. Ideally, all the scientists working on the alliance project would attend, as would support personnel and managers who will need to interact with the partner over the course of the alliance. It is also important to include higher-level research managers and executives from both companies so they can hear, and clarify (if needed), the goals and timing of the project.

Senior leaders are there to set the stage, to show how well they already work together and to model effective communication and decision-making. In addition to formal presentations, senior leaders might share their stories about getting this far (i.e., to a signed deal), illustrating that progress is not linear. The process of reaching contractual agreement is characterized by "bumps" along the way (prefiguring the inevitable lumps and bumps over the alliance life cycle). At one initial meeting, the counsel for a biotechnology company recounted the following story:

> Our president put together a package that asked for the sun, the moon, and the stars. Of course, they balked. Everyone walked away, although both firms continued to work on the experiments. About 6 months later, the other company's internal program blows up—they discover a ligand, but the compound is owned by [X]! So, they came back to us. This time, our president asked for smaller sun, moon, and stars, and after a lot of frank discussions, she got them. We finally figured out how to work together.

Scientists are there to present, learn, and share. Key scientists from both partners should describe relevant aspects of their work and answer questions, because each side needs to learn what the other is doing. Also, enough senior scientists from the pharmaceutical company should attend, or some fine points in the research plan may be missed. What should emerge from these presentations and discussions is a clear sense that the collaboration, as described by one of our interviewees, is "a really important thing":

> We should be collaborating because there is synergy. The only reasons to work together are to get synergy, to gain skills, to increase competence (biotechnology company CEO).

People on the biotechnology side must ensure that those who attend do not give the impression of arrogance:

> Scientists from biotechnology companies can be very arrogant. They come into the meeting and insist that the work be done their way, this many FTEs and so on. No: it does not have to be done *their way*—it has to be done *the team's way*. People on both sides have to set up a plan in which who does what is based on the right set of skills, decided by the team, by the people in the collaborating group.

> Of course, not every lead scientist is arrogant! But, they rarely see the link between project success and company viability (pharmaceutical executive).

Time, Place, Duration

The best time to hold the first formal meeting is as soon as senior-level people are willing. The sooner this first meeting is held, the more impact it will have in terms of how work is done and how people feel about working together. In many ways, the meeting should be held as an alliance kick-off event, a time to celebrate the deal and set the work in motion.

Where should it be held? As a rule of thumb, the meeting should not occur at either partner's facilities. First, getting to know each other is more easily accomplished when both parties come together as equals and neither has a "territorial" advantage. Second, the meeting is vital, so it should be difficult for people to leave and go to their office or laboratory to transact other business.

There is likely to be greater commitment to participating in the first meeting if it is held in a pleasant (not lavish) location, where people can be comfortable and can enjoy themselves. The facility should provide opportunities for people to spend time together outside the formal meeting hours, eating, talking, socializing, and so on.

Given these recommendations, it might appear that this first meeting is taking time away from getting the real work done. In fact, this meeting *is* real work. As we noted above, there will be high levels of uncertainty and equivocality about the project and about collaborating that must be addressed effectively. One of the best arrangements is to begin the meeting on one afternoon, have dinner together, continue through the next day, and then finish up on the following morning. Such a schedule allows time for dealing with a full range of scientific and technical matters that are important and allows time for build-ing relationships.

CLOSING THOUGHTS

If we begin with the premise that both partners want the alliance to be successful, then both partners should take some responsibility for the first meeting. Because much of the day-to-day management of the relationship will fall on the shoulders of people on the biotechnology side, it makes sense that they will bear primary responsibility for the meeting. Of course, it would be even better if both sides were involved in planning, organizing, and leading the events and activities.

Oftentimes, it makes sense to have the person in the Commitment Champion role take on the responsibility for the meeting, delegating as needed and working with the Science and Technology Champion and Process Champion. The objective is to be sure that scientific and interpersonal preparations are completed.

There are numerous logistical details to be worked out, but these are not our concern here. Rather, we want to emphasize the following:

- Senior people need to be committed to the first formal meeting. They need to clear their calendars, and they need to come to some agreement and clarity about their roles at the meeting.
- Scientific and business presentations must be prepared and rehearsed. Ground rules should be agreed upon between the partners. If each partner has a "meeting project manager," that person should explain

the purpose of the meeting, clarify expectations, and identify some ground rules.

- Finally, if meeting planners feel a need for assistance, an outside consultant could be brought in to help design activities for people to get to know each other and learn about working together. If possible, the alliance managers should actually "run" the exercises, in order to give them greater credibility to an audience not known for appreciating the interpersonal aspects of work and relationships. (The consultant should, if possible, be on hand during the meeting to help them.)

These are all simple steps. But, by taking them, partners embark on a trajectory that will maximize the likelihood of effectiveness and, in the end, create benefit for both parties.

Think differently . . .

Lead differently . . .

Make alliances work.

PART IV

THE ALLIANCE LIFE CYCLE: LEADING DIFFERENTLY OVER TIME

Beginning with Chapter 5, we have essentially been following a single alliance (albeit a hypothetical one). Chapters 5 through 7 dealt with preparedness, beginning with the biotechnology organization, moving through due diligence on the partner, and concluding with the first meetings.

Chapters 8 through 10 continue the chronology and map-specific portions of the collaborative work. Like many other phenomena, alliances can be depicted as having a life cycle (the stages are called *emergence, growth, maturity,* and *completion*). Chapter 8 examines dynamics, leadership issues, and leadership roles in the first (emergence) stage, which concludes with accomplishment of the first major milestone. Chapter 9 addresses predictable issues of the growth and maturity stages, particularly interpersonal and process issues that surface once the milestone is achieved and demand leadership attention. Chapter 10 discusses the desired-for completion of the project, as well as the undesired result of termination.

We use a life-cycle model, because the tensions, issues, and challenges that occur in each stage are generally predictable and have implications for alliance dynamics. What can be predicted, of course, can be anticipated, planned for, and intelligently managed. Moreover, different skills and abilities are needed by leaders at different stages in the cycle.

The reader will also note the importance of the two concepts defined in Part III: uncertainty and equivocality. In fact, levels of these will vary predictably over the stages of the life cycle, again with implications for alliance dynamics. Understanding these implications and dealing with them effectively are crucial to alliance effectiveness and alliance success.

CHAPTER 8

TO THE FIRST MILESTONE

After the first formal meeting, team members should have a better understanding of the alliance and a better appreciation of players on the "other side." What will now loom large in the project timeline is the first milestone. For most biotechnology companies, receiving a sizable payment is contingent upon reaching that milestone, thus accounting for high levels of anxiety, but the milestone is important for a number of other reasons. In this chapter we describe the first (emergence) stage of the collaborative effort and address leadership issues involved in accomplishing the initial contract objectives.

THE CHALLENGING FIRST STAGE

The first formal meeting signals that the alliance has officially begun. When they return to their respective firms, people will have an appreciation of the different players and of ways to improve working together to accomplish the lofty alliance goals ("enable us [big Pharma] to enter a new therapeutic area;" "provide us [Biotech] with a better knowledge of clinical considerations in [X] therapy").

But, these ultimate alliance goals are not what members of the team will be working toward during the initial stage. Immediately, they must

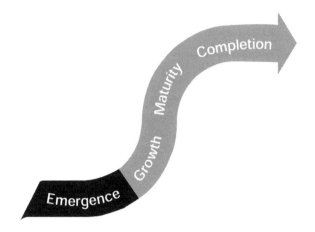

concentrate their efforts on reaching the first milestone. Note that we are not addressing questions about how, and by whom, milestones are set. Rather, we begin after a deal has been signed and the milestone has been identified as part of that process. Of course, realistic milestones and realistic timeframes will make alliance effectiveness and success more likely.

Moving toward the milestone is a difficult period for leaders, managers, and scientists. People will be anxious and tense during this stage of the collaboration, because of what hangs in the balance:

- *Revenues for the Biotechnology Company.* Earlier, we cited a press release on the failed Lucida–Pharma Sciences alliance that stated "Lucida [was] no longer eligible for most of the $80 million in milestone payments." Biotechnology firms flourish or falter according to how well the team meets the first milestone.

- *The Reputation of Alliance Advocates.* In both the big pharmaceutical and the small biotechnology partner firms, individuals have advocated for this collaboration (so-called *champions* of the alliance). Their reputations can rise or fall according to how well the team meets the first milestone.

- *Potential Competitive Advantage.* Based on organizational due diligence, leaders have determined that the collaboration adds strategic value to each company. Desired competitive advantage for both companies will be increased or diminished according to how well the team meets the first milestone. Although big Pharma does not need this particular alliance to succeed in the short term, there must have been perceived strategic benefit for the company to engage in the collaboration.

* *Maintaining Senior Managements' Interest in the Alliance.* Efficient attainment of the first milestone fosters confidence among the senior managers. Unwarranted delays have the opposite effect.

This tension was noted by several of our interviewees from biotechnology companies:

> People in big pharma tend to be skeptical of an alliance, until the first milestone is met. This is especially true if this is the big company's first experience with our firm, or with our technology.

So much is riding on the accomplishment of the milestone objectives that everyone will feel pressure. Tension will characterize the first stage of an alliance, but this is neither an indication of "fatal" problems nor a condition to be avoided. We want to emphasize that the unease of this stage can be put into a context that makes it more understandable, predictable, and manageable. This context is the *life cycle*, which we address in the following section. Later, we discuss the sensitive topic of the *difficult scientist* as this relates to achieving the first milestone. We conclude with additional tasks and activities that biotechnology leaders must oversee, if not carry out themselves. To ensure that nothing is overlooked, we have organized the material according to the four alliance leadership roles (sets of interrelated activities, behaviors, and tasks), although one individual may occupy more than a single role.

THE LIFE CYCLE

The S-shaped curve of change in some parameter over time is a very useful depiction of a variety of natural phenomena, such as growth of populations and spread of infectious disease. Life cycles can also be used to depict artefactual phenomena, such as pattern of scientific citations, diffusion of innovation, use of discovery technologies, accomplishment of project tasks, and completion of biotechnology alliances.

Life cycles vary in duration. Some are relatively long; others are relatively short. In certain industries, an alliance life cycle is of relatively long duration. Car manufacturers may have alliances of many years with parts suppliers. In other industries, the alliance life cycle may be much shorter. Biotechnology alliances have a relatively short duration, some even shorter than 3 years. Although a few have lasted a decade or more, discovery alliances between pharmaceutical and biotechnology firms often run less than 1 year.[1] Strategic alliances in the fast-moving biomedical sciences present special leadership challenges because the life cycle is of short duration.

A life cycle is typically divided into four stages—the four parts of the S-shaped curve—termed emergence, growth, maturity, and decline. (These stages may also be called formation, early/late buildup, main, and completion or termination.) Although stages are depicted as of equal length, this is just an idealized representation. The first stage (emergence) may be much longer than any other stage in reality. Or, the rising part of the curve (growth) may be the longest; or, the flattening out of the curve (maturity).

Within the same industry, the length of stages of relevant discovery technologies may differ across sectors. As we discussed earlier, medicinal chemistry emerged as a discovery technology in the nineteenth century and, over the next 30 or so years, was incorporated into the evolving pharmaceutical industry. Today, medicinal chemistry is in the mature stage of the life cycle, based on the metrics of "numbers of pharmaceutical companies founded" and "year founded." On the other hand, biotechnology emerged in the 1970s as a discovery technology and was rapidly incorporated into the evolving biotechnology sector. The emergence stage of each discovery technology was, thus, of markedly different duration. Today, biotechnology is in the growth stage of the life cycle, based on the metrics of "numbers of biotechnology companies founded" and "year founded."

Life-cycle stages of organizational activities (like alliances) or technologies (medicinal chemistry and biotechnology) are also artifactual. They are social conventions, based on perceptions of the people involved. Thus, what is perceived to be in the emergence (i.e., earliest) stage for one group of people may be in the growth stage (later stage) for another, because of differences in experience between the two groups. An important finding of our study of biotechnology alliances is that there is an *unavoidable asynchrony* in experience between the partners.

Using life-cycle terms, there are life-cycle stages out of synch at the time of signing the alliance deal. From the perspective of people on the biotechnology side, their technology (the means by which the work of the alliance is

[1]Based on Recombinant Capital database (recap.com).

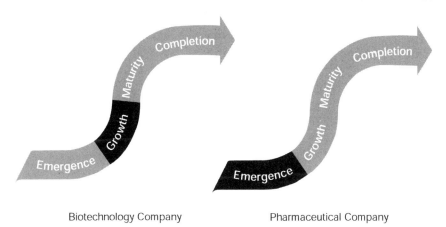

Biotechnology Company   Pharmaceutical Company

Two life cycle stages out of synch.

to be accomplished) is at a later stage than it is for those on the pharmaceutical side. Simply stated, when the biotechnology firm was founded, there was a "concept" judged to be worth funding but there was little of interest to a pharmaceutical partner. Many biotechnology executives with whom we spoke described a 2-year interval "in the founder's lab," before potential value for a "concept" could be demonstrated to any partner's satisfaction.

When value can be shown, the "concept" (the technology) will, by definition, be perceived to be in a later stage of its life cycle by the biotechnology scientists who are experienced in its use. From the vantage of the pharmaceutical scientists, however, the technology ("concept") will be perceived to be in the emergence stage when the alliance is signed. We have found that this asynchrony, stemming from differences in experience and perspectives of the collaborating scientists, adds to the leadership challenges of the first stage.

The utility of the life-cycle model for those leading and working in biotechnology alliances is this: In each stage, there are different tensions, issues, and challenges that are generally predictable and have implications for alliance dynamics. What can be predicted, of course, can be anticipated, planned for, and intelligently managed.[2] Moreover, leaders must employ different skills and abilities at different stages in the cycle.

[2]Galbraith, J. *Designing Complex Organizations*, Reading, MA: Addison-Wesley, 1973. See also *Competing with Flexible Lateral Organizations*, 2nd ed., Reading MA: Addison-Wesley, 1994.

The emergence stage of the alliance life cycle, the time from the kick-off meeting to the first milestone, is predictably characterized by tension arising from the fact that so much is riding on successful accomplishment of the milestone. Everyone on the biotechnology side must understand that the pressure is unavoidable, but it cannot be allowed to cloud judgment or mar collegial relationships. Leadership must ensure that work is undertaken in systematic fashion, that sufficient time is spent reflecting on problems that will invariably arise in the work, and that problems are calmly discussed with the partner.

Leadership must also deal with the challenge that arises from the differences in experience of the two groups of scientists (*life-cycle stage asynchrony*). We address this and other challenges under the headings of task uncertainty, task equivocality, and scientific and technical tacit knowledge, below.

Task Uncertainty

The technical work of the alliance is embedded in an information environment that contains cues, hints, data, signals, and other intangible resources required to make progress. The gap between what collaborators need to bring the alliance to completion and the amount they have at any given time is defined as uncertainty.[3] As a rule, this gap will be widest (the level of task uncertainty will be highest) during the first stage of the alliance life cycle. *A leadership goal is to reduce this level efficiently.* Reaching the first milestone within the specified period depends (among other things) on how efficiently (quickly) uncertainty can be reduced.

In the first formal meeting, whether they realize it or not, all attendees are engaged in the process of reducing uncertainty by seeking, disseminating, and sharing information. Presentations (by company leaders) of strategic objectives, of the likely fit of the invention into the large company's portfolio, and of the desired competitive advantage provided by the outcome help to reduce uncertainty about the alliance generally.

Although there will be presentations on the science and technology during the first meeting, the level of task uncertainty (as opposed to what we might call "alliance uncertainty") can only be reduced by experimentation and the dissemination of new results. But, there are differences in individual and collective experience on either "side" of the collaboration (*life-cycle stage asynchrony*). This may lead to different perceptions of what

---

[3]Galbraith, 1973 (see footnote 2).

Task uncertainty.

further experimentation is needed. Thus, what is less uncertain to some may be more uncertain to others.

At the start, pharmaceutical scientists are likely to perceive a higher level of uncertainty about the task (the science and technology), for the simple reason that experience, information, and knowledge reside with others in the biotechnology firm. Pharmaceutical scientists must reduce uncertainty and "catch up" by reviewing data and results and by replicating some of the experimental work. Because of the differences in experience, however, initial replication of a biotechnology experiment is likely to require more than a single attempt by the large partner. Even the definition of "replication" can differ between the two partners, so that very explicit questions should be asked (e.g., "Are 4 days sufficient for the experiment or should 8 be used?" "Will sampling every 30 minutes be needed or will sampling every 2 hours be enough?" and so on).

One biotechnology scientist commented:

> What happens in all discovery alliances is that you come up with data. Then, you sit down with the partner, who tries to replicate your data. But, they can't succeed!

Then the partner says: "Hey, wait a minute. What did I pay for?" That happens in virtually every relationship. By the third time, the data begin to look right.

Leadership may not be able to ensure that a single experiment reproduces the desired results, but three or more experiments may be unnecessary. Reaching agreement on experimental design could be crucial at this point. Skimpy design versus "overkill" in design needs to be balanced (and agreed upon). To reduce uncertainty efficiently, leaders from the biotechnology firm should plan to hold several meetings for intensive discussion of the experimental conditions. Presentations and backup information should provide sufficient detail that much of the experience, information, and knowledge residing on the biotechnology side can be shared. Such sharing also helps to rectify the asynchrony of life-cycle stages.

In addition, the design of the project (including resources of time and cost) must allow for interchange of scientists. Some time must be built into this early stage, to permit relevant pharmaceutical scientists to work side by side with their biotechnology colleagues, at either site. This latter tactic is also indicated because of the high level of task equivocality and the existence of key, tacit knowledge on the biotechnology side during the emergence stage of the alliance, as discussed below.

Task Equivocality

The information environment of the alliance task will also be fuzzy, ambiguous, or equivocal.[4] Such ambiguity is commonplace and reflects a gap. However, this gap represents the difference between the level of shared interpretation, meaning, and understanding needed to accomplish the work and the level the team shares at any given time.[5] As a rule, the gap will be widest (the level of task equivocality will be highest) during the first stage of the alliance life cycle. *A leadership goal is to reduce this level efficiently.* Reaching the first milestone within the specified period depends (among other factors) on how efficiently (quickly) equivocality can be reduced.

[4]Daft, R., and Lengel, R. *Information Richness: A New Approach to Managerial Behavior and Organization Design*, in Staw and Cummings (Eds.), Research in Organizational Behavior, Greenwich, CT: JAI Press, 1984.

[5]Stork, D., and Sapienza, A. Uncertainty and Equivocality in Projects: Managing Their Implications for the Project Team, *Engineering Management Journal*, Vol. 7, No. 3, September 1995.

When the level of *uncertainty* is high, more information must be sought and exchanged. But, when the level of *equivocality* is high, the process is not so straightforward. After all, we are dealing with interpretation as opposed to information. When equivocality is high, it is not clear what kinds of information would be helpful or even what kinds of questions should be asked. To reduce equivocality, people must interact and share their different values, perspectives, and interpretations (not information) and resolve their conflicting views.

Uncertainty and equivocality are social constructs rather than attributes of any particular task or technology. Thus, members of the alliance team from the biotechnology side may experience levels of uncertainty and equivocality that are different from those experienced by their counterparts on the pharmaceutical side, even though they are trying to accomplish the same task or are working with the same technology. For example, the asynchrony of life-cycle stages that we described earlier is likely to result in higher perceived equivocality by pharmaceutical team members, as revealed by one interviewee:

> During the first meeting we pay close attention to the questions that scientists from big pharma ask. Often, the questions surprise us, but we have come to realize that people ask these questions because they have a different perception of reality that they believe is true. We got over being irritated by this! (biotechnology project manager).

To reduce the level of task equivocality, team members must interact with one another directly, until they come to share the same perceptions of "reality." To reduce the level of task uncertainty, information gathered or created by any one member can be shared with other members. Unlike equivocality, which is reduced by the creation of shared meaning, uncertainty can be reduced by simply transmitting information clearly, so that it is perceived and understood as it is meant to be. Reducing equivocality, however, requires face-to-face discussion among those concerned, until shared agreement about an *interpretation* is reached.

Not surprisingly, given their different training and experience, scientists from each firm will have different interpretations of their collaborative work. When experimental results from biotechnology efforts are presented by team members, they are likely to be interpreted differently by scientists from the large partner. Similarly, results from the large partner's attempts at replication are likely to be interpreted differently by biotechnology scientists.

Because both task uncertainty and task equivocality are high, there should be more face-to-face meetings of the team during the first stage of the alliance than at any other stage. On this point we found agreement, from both the pharmaceutical and the biotechnology perspectives.

> . . . There has to be constant interaction with the biotechnology scientists. Quarterly meetings will not do it; you sometimes need weekly meetings until you reach the first milestone. At least, the frequency needs to be such that your discussions result in agreement about the results and how to proceed. That's based on understanding the body language of your partners, which is an outcome of the informal interactions you have had in the team formation period (pharmaceutical vice president).

> . . . I have learned that I cannot stay too much on top of the work, or have too many meetings, while we're working towards the first milestone. I don't need all the scientific details, because I know where morale is or is not by my people's body language, and good morale means the work is going well. Poor morale means trouble. What is hard for me is finding out how well the work is going in the partner firm. That's why we have so many meetings! I need to see their body language as well (biotechnology project manager).

As we explain below, much of the scientific and technical knowledge on the biotechnology side is *tacit*. Thus, another leadership challenge in this stage of the alliance is to ensure that key, tacit knowledge becomes available to the team.

Tacit Knowledge

Knowledge exists as two different types, tacit and explicit, and in two domains, individual and collective (or shared).[6] In the first stage of the alliance, we can state the following as a general rule:

Most of the scientific and technical knowledge then existing will be of the tacit *type and in the* individual *domain on the* biotechnology side.

---

[6]Nonaka, I., and Takeuchi, H. *The Knowledge-Creating Company*, New York: Oxford University Press, 1995.

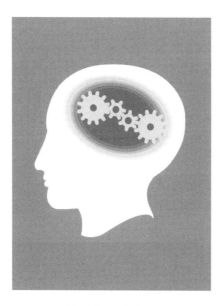

Tacit knowledge.

The word "tacit" is derived from the Latin verb *to be silent*. Tacit knowledge is that which is understood but is not communicated, because it is without words or speech. Some authors have defined tacit knowledge as "person embodied," "on the hoof" so to speak.[7] For example, a biotechnology scientist may talk about her work and explain her methods in great detail to a partner scientist, but, by definition, she will not be able to communicate her tacit knowledge. Sharing tacit knowledge requires that one scientist *rub shoulders with* (i.e., work side-by-side) the scientist who possesses the tacit knowledge.

Tacit knowledge can be said to be imparted by osmosis—that is, by observing, copying, and trying an activity in concert with someone who possesses the tacit knowledge. Of course, the reality of today is that alliance work is not accomplished at a single location. Thus, time and resources must be devoted to helping scientists work "at the same bench," even if it may seem as though they are being taken away from the collaborative work. In fact, only by working, reviewing results together, discussing experimental conditions and outcomes, setting up the experiments together, and so on, can tacit knowledge be shared.

[7]Faulkner, W., and Senker, J. *Knowledge Frontiers*, Oxford, UK: Oxford University Press, 1995.

We emphasize tacit knowledge in this book, because (without exaggeration) tacit knowledge is *the* reason for the alliance. If all the scientific and technical knowledge possessed by the biotechnology scientists were explicit, there would be no need for the pharmaceutical company to collaborate. Pharmaceutical scientists would only require the "how to" report or journal articles. But, when the frontiers of science are advancing rapidly, much of the knowledge being generated will be tacit. That is why biotechnology and pharmaceutical scientists have to work together, literally, in order to share it.

In contrast to tacit knowledge, explicit knowledge is that which can be communicated. As we suggested earlier, experts possessing explicit knowledge give lectures and write articles and, thus, can communicate that

Explicit knowledge.

knowledge to other individuals or groups. In the collective or shared domain, explicit knowledge is what usually fills libraries, departmental archives, and databases on file servers. It is available and transmittable, if one knows where to look.

In the first stage of the alliance, a key task of biotechnology leaders is to ensure that the scientific and technical knowledge that is individual and tacit becomes available to the alliance team. Note that we have not described the challenge as facilitating the transformation of *tacit* knowledge to *explicit* knowledge. Our reasons: the long time it would take to do this, and the high probability that the outcome (that which became explicit) would be incomplete. Experts, or those who embody particular tacit knowledge, gained their expertise over a period of years. Attempting to categorize, enumerate, and describe that expertise is a lengthy process.

More importantly, although experts may believe that they "know what they know" and can explain their expertise to another, much of their tacit knowledge is that which they "do not know that they know." It is only when the individual scientists spend time with each other that there is an opportunity for the tacit to be "absorbed." Over successive stages of the alliance life cycle, the shared tacit knowledge will eventually become explicit. Explicit knowledge, after all, is what signifies that the contractual requirements of the alliance are being met.

In the context of the life-cycle model and its attendant issues (especially uncertainty and equivocality), we can summarize that effective leadership during the first stage of the alliance requires the following:

- Understanding and managing the tension and unease associated with meeting the critical first milestone
- Facilitating the efficient reduction of scientific and technical uncertainty, by facilitating the communication and sharing of explicit knowledge
- Facilitating the efficient reduction of scientific and technical equivocality, by holding sufficient face-to-face meetings that people come to shared agreement on the interpretation of experimental results
- Enabling the processes by which individual tacit knowledge from the biotechnology side becomes available to the whole team, by putting people together for scientific collaboration.

In the next section, we address a subject both sensitive and crucial, particularly in light of the first milestone and the alliance goal of creating new knowledge: the *difficult scientist*. Some experts may be difficult to work with in a team environment. However, they have to be managed effectively, so that their tacit and explicit knowledge becomes available to the whole team.

THE DIFFICULT SCIENTIST

Let us state upfront that not every expert is difficult to manage in situations calling for teamwork. But, there are enough examples of difficult experts for us to devote a short section to this topic. Our Lucida case study presented a difficult scientist, "Will O'Brien," who was described as follows:

> O'Brien is very smart scientifically, but he is not managerially inclined! He believes that, if you're the boss, you tell people what to do and they do it. He's one of my best friends, and we've worked together for years. But, he has said to me: "I'm the only intelligent scientist in this company. Everybody else is a blithering idiot!" Now, I'm a scientist, and I'm his friend, but he says this with genuine sincerity.

The difficult scientist.

Many experts, like O'Brien, do best as individual contributors. Their training, education, and experience have taught them to be individualists. They are generally not interested in leading others (apart from "telling people what to do"), nor are they interested in how the collaboration is doing or how the company is doing. They are interested in how well they are doing, scientifically and technically. Their personal achievement and scientific reputation are their primary concerns. One experienced leader described these experts as "self-starters, . . . 'noncompliant' [with regard] to formal organizational needs, . . . hold[ing] strong convictions, forcefully expressed."[8]

If not managed effectively, the difficult scientist may jeopardize the accomplishment of the first milestone, as happened in Lucida:

> . . . O'Brien believed he should be in charge of the program, but Pitchly [the CEO] and Rosenbloom [the Chief Scientific Officer] did not want to put him in charge. . . . Of course, during the joint team meetings, O'Brien was publicly dysfunctional.

> We certainly did not come across as a coherent organization. Pharma wanted certain data. O'Brien had the data, but because he was not in charge he was not going to give the data to the project manager.

As the Lucida situation showed, although difficult scientists may function best as individual contributors, they are crucial to the scientific and technical progress of the biotechnology collaboration. Their tacit knowledge is extremely valuable; yet, working side by side with them may try others' patience and forbearance to the breaking point. Certainly, the difficult scientist will strain the leadership capabilities of those in charge and, in the end, can profoundly affect the progress, even the success, of the alliances. One of the Lucida interviewees talked about lead scientists (often the difficult ones):

> These lead scientists don't necessarily fear project failure, because they are buffered by the interfacing people. They may not always see the connection between their behavior and how effectively or not the work progresses.

> Such people are often the reason the biotech company was funded in the first place. They are revered within the organization, and few want to challenge

---

[8]Fitzgerald, J. D., cited in Thorne, 1992 (see reference in Chapter 2).

them. Sometimes a biotech company will hire someone to manage the collaboration. The job is bigger than that, however! The job really involves managing the lead scientist.

Again, it is the interface that is complicated. The lead scientist probably feels that she or he should be leading the collaboration and may not appreciate having a "lesser light" in charge of the project.

Assuming that the leader has tried to win the person over to the team effort (e.g., by highlighting the scientific challenge and by addressing the "what's in it for me?" issue outright) and failed, we suggest two tactics. First, we recommend that a difficult scientist be treated as a virtual member of the alliance team. This scientist should always be informed of the scientific and technical issues but not necessarily included in meetings that address nonscientific or nontechnical topics (collaborative processes, alliance strategies, and so on). When the scientist must attend a meeting, he or she should come for that particular scientific agenda item. All team members should be prepared with their questions and issues; and, as far as possible, at least a short list of these should be given to the expert ahead of time. The objective is to minimize the number of "surprises" for this scientist, thus reducing the possibilities for defensive behavior. The scientist should be considered an invited expert and encouraged to teach, show, or demonstrate, rather than discuss. During the meeting, individuals should be encouraged to seek clarification from the expert (who will usually enjoy this role). Afterwards, team members can converse and discuss what was presented and engage in lively, challenging conversations without provoking "dysfunctional behavior."

Virtual team member.

The second tactic is to identify a scientist who will be responsible for "rubbing shoulders" with the difficult scientist, in order to gain the needed tacit knowledge. This scientist should be easygoing, not easily intimidated or annoyed by the expert, competent in the appropriate scientific and technical arenas, and able to communicate what he or she learns to the other members of the team. In short, this person will be a physical link between the virtual team member and the "real" team members.

As emphasized by the above-cited interviewee, people in alliance leadership roles must take responsibility for managing the difficult scientist. That was not the case in the Lucida–Pharma Sciences alliance, before the vice president of research (Santoro) was hired. As Santoro discovered, O'Brien was described by the company president as a difficult person and someone Santoro would have to "deal with." Santoro assumed that the Chief Scientific Officer could not, or would not, manage O'Brien, and the president appeared very unwilling to deal with him.

Having said all that, we want to remind the reader that, challenging though they may be, these people are good for the organization:

> There are delicate flowers of egos in both biotechnology and pharmaceutical companies. Every organization has them and every organization needs people like that. You just have to know how to handle them. You need them because you're not producing widgets—you're producing things that never existed before. You get strange people and prima donnas in this business. You may get more in discovery, but you need them! You have to have them, to create what did not exist (biotechnology project manager).

In the concluding section of the chapter, we review other tasks and activities that leaders from the biotechnology side must oversee, if not carry out themselves, to reach the first milestone.

LEADERSHIP ROLES IN THE EMERGENCE STAGE

There are four critical leadership roles in the biotechnology firm, as we described earlier:

- Commitment Champion (senior executive with general and ongoing alliance responsibility)
- Process Champion (leader with specific responsibility for keeping the alliance processes running smoothly)

- Science and Technology Champion (subject matter expert responsible for scientific and technological progress of the effort)
- Alliance Team Member (responsible for leading parts of the collaboration in specific areas of expertise).

We stressed the difference between roles and people. This distinction is still important, but for the sake of easier writing and reading, we will refer to "champions" when, in fact, we mean "championship roles." This is simply a convenient shorthand; we do not imply a single individual when we say: "The process champion should. . . ." One individual may carry out the activities and tasks of more than one role. Or, a single role may be shared by more than one individual.

Each of the four alliance leadership roles is crucial to the success of the alliance. Fundamentally, each of the roles is focused on *getting there*, with "there" at this stage of the alliance life cycle being the first milestone. Before we identify some of the critical activities and tasks of each role, we want to emphasize two general issues that are relevant for all leadership roles.

First, levels of both uncertainty and equivocality are high during the first stage of the alliance effort. Thus, leaders must be able to distinguish between those problems for which simple information processing is needed versus those that require social interaction and face-to-face communication to resolve. Second, at this stage, the relevant alliance knowledge is tacit and exists largely at the individual level. To reach the crucial first milestone, this knowledge must be shared with all team members, and at least some of it must become explicit.

Commitment Champion

The Commitment Champion will be somewhat less involved at this time than he or she was earlier and will be later, when some of the natural alliance "endorphins" begin to subside. Before and through the first meeting, the Commitment Champion was garnering support, organizing resources, and building a base of commitment for the long haul. In the period between the first meeting and accomplishing the first milestone, the Commitment Champion needs to build up his/her own physical and psychological reserves for the future.

During the emergence stage, the Commitment Champion should focus on developing and sustaining relationships. When increased involvement and commitment from the partner are required later, as they invariably will be, good, collegial relationships will be critical. When there are good relationships, telephone calls and e-mails will be answered in a timely fashion. People will find room in their schedules to meet. They will also give more (rather than less) benefit of the doubt when progress appears shaky. In other words, they want to win by having this work move forward and eventually succeed.

Although developing and sustaining relationships may not appear to be *real work*, we can assure the reader that this is indeed work on the critical path. How relationships are built and maintained is too context-specific to be addressed in this book. Suffice it to say that (at a minimum) periodic informal, candid, face-to-face meetings on alliance progress, prompt and open response to questions, discussion of critical issues, and so on, are an important task of the Commitment Champion role.

Process Champion

Earlier, we noted that the Process Champion keeps the alliance invigorated, focused, and functioning smoothly. During the emergence stage of the life cycle, the Process Champion must concentrate on removing impediments to team efforts and facilitating collaborative activities. It is the Process Champion who helps smooth the bumps in this part of the road. Also, because tacit knowledge becomes available to the team mainly through the "rubbing of shoulders," it is the Process Champion who must enable the interchange of scientists and shared working arrangements.

We realize that smoothing bumps and exchanging scientists are easier said than done. Issues of territoriality and ownership often get in the way of genuine collaboration. Even those scientists who have little difficulty in passing on information and sharing data and results may bristle at some of the expectations of an alliance. Moreover, companies have policies, and big companies may have many policies that produce conflicting priorities for an individual and that constrain travel and time spent on the alliance project. If the Commitment Champion has developed good relationships, however, it should be possible to get agreement about travel and time. Effective leaders from the biotechnology side will have raised the possibility of travel and time before permission is required. The resources, ideally, will have been incorporated into the project plan, and the time and cost will have been included in the alliance total.

Bringing about the ethos and practice of collaboration, in ways that ensure the availability of tacit knowledge to the team, requires that the Process Champion be sensitive to organizational culture, to issues of ego and autonomy (and "not invented here"), as well as to the expected interpersonal dynamics. But, when "cowboys" in the laboratory can really work together, in a sharing, common-goal environment, alliance progress will be faster.

Science and Technology Champion

Of the four roles, we expect the Science and Technology Champion to be exceptionally busy during the emergence stage of the life cycle. As a general rule, some scientists on both sides will be newcomers to some aspects of the science and technology. The Science and Technology Champion must understand that knowledge, understanding, and experience will vary and that rather substantial teaching responsibilities are part of this role.

This champion has much to learn as well, some of which should precede the teaching responsibilities, some of which may run in parallel. The Science and Technology Champion must understand as much as possible about the science and technology issues, questions, steps, and setbacks on both sides. Reducing task uncertainty and task equivocality are critical responsibilities. Ensuring that the asynchrony of life-cycle stages is eradicated is crucial. Thus, we believe that keeping abreast at the requisite level of detail will tax this champion's knowledge acquisition and interpretation skills.

Although the Science and Technology Champion role is focused on science and technology, interpersonal skills are also important. Mutual respect and trust facilitate the sharing of important information. Without open relationships (both within and between groups), information sharing will suffer and the more "problematic" information may not be shared. Using earlier terms, uncertainty will not be reduced efficiently, and the project timeline will suffer.

In an alliance, withholding information can derail an otherwise promising project. Let us remember that, at Lucida, certain key information was not shared either within the company or between the Lucida and Pharma Sciences groups:

When results came in, they were ambiguous. It took weeks before we could understand what was going on. Turns out there were longtime scientists here who were fonts of information. But, they would sit in a meeting with Pharma people and bring something up and we would say to ourselves: "Oops! I never heard that before!"

Mutual respect and trust also facilitate the reduction of equivocality. To achieve a shared interpretation of, for example, experimental results, people must believe that their own interpretation will not be derided or scathingly judged. Interpretations will be respected, if relationships are built on respect.

Alliance Team Members

For Alliance Team Members, reaching the milestone requires dogged determination and focus. Although tempting, there can be no excursions into "interesting" science, no dramas, no tries for brilliance in tangential areas. The emergence stage of the life cycle is the time to stay on the straight and narrow, both scientifically and interpersonally. Egos and individual preferences have to be set aside for team goals rather than individual glory.

Very briefly, the primary task of each Alliance Team Member is to engage in competent, clean, reproducible, and well-documented scientific experiments, the results of which are willingly shared, discussed, challenged, and critiqued. It may be tempting to consider shortcuts, to neglect the laboratory notebook, or to repeat experiments that "didn't work" without telling anyone. None of this is acceptable. The road to the milestone is paved with discipline, integrity, hard science, and critical thinking.

CLOSING THOUGHTS

Although they may seem to be heroic assumptions, we close this chapter confident that the milestone objectives will be met, that money will be paid to the biotechnology firm, and that the "burn rate" will slow to a more manageable pace. Our focus can now shift to the different, but expected, leadership challenges of the growth and maturity stages of the alliance life cycle.

Think differently . . .
        Lead differently . . .
                Make alliances work.

CHAPTER 9

MANAGING GROWTH AND MATURITY

For the alliance team, accomplishing the first milestone is a time for celebration. It is also a time for leaders to examine possible consequences of the intense pre-milestone activities. We expect conflicts and problematic dynamics to have been set aside; certain efforts to have been postponed; and some project tasks to be behind schedule. This chapter discusses both post-milestone collaborating and the particular leadership challenges of the growth and maturity stages of the alliance life cycle.

AFTER THE MILESTONE

First, celebrate. . . . When the milestone is achieved, the alliance team *should* breathe a sigh of relief and celebrate. This is a time to anticipate success.

It is also a time for leaders to examine the collaboration thus far. We expect there to be "fallout" from the intense pre-milestone activities. Despite the best intentions, disagreements and conflicts may have been swept under the rug in the focused effort to achieve the initial contract objectives. Because of conflicting priorities, some tasks will have been postponed, putting the project behind schedule. People will begin to feel pressured (again), as target dates slip.

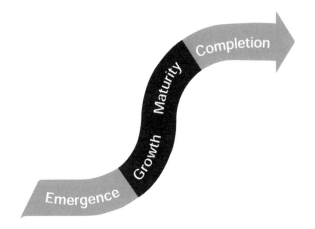

The success of the alliance depends not only on finding good solutions to, for example, scientific and technical problems associated with the first milestone but also on developing and sustaining good relationships among scientists and leaders in the two companies. At this point in the alliance effort, leaders should focus on the process of collaborating. Again, the larger burden of responsibility rests on people in the biotechnology firm:

> Who bears more responsibility for success? Of course, the biotech leadership. If the alliance goes belly up, it is really all over for everyone. In big Pharma, they know they will have a job when they come to work in the morning, but you do not know that in biotech. In a small biotech company, you have to be a better manager of the partnership (biotechnology project manager).

We understand that the general processes of collaborating—interpersonal dynamics, communication, conflict resolution, and so on—may not get the consideration they deserve during the emergence stage of the alliance. During that time, people's attention will be diverted by technical issues. Attention will also be diverted by the positive energy of getting started and by the anxiety of attaining the first milestone. The leadership challenges that surface in the immediate post-milestone period are, to a large extent, the expected consequences of what has been put on the back burner. Many of the potentially difficult human resource issues were probably ignored or dealt with inadequately.

Compared with a research team in a single organization, alliance teams face more complicated and problematic interpersonal dynamics, for several reasons:

- The alliance team is virtual. No laboratory space is shared over the entire duration; rather, experiments are performed in parallel as well as sequentially (and only rarely together, shoulder-to-shoulder). Work is primarily accomplished, or not, at team meetings.
- The relevant scientific and technical frontiers of knowledge are moving rapidly, and the alliance has a relatively short life cycle. It is difficult in a short period to "get it right" (both the science and the interpersonal dynamics).
- There are not "two elephants in this ring." The large pharmaceutical company is undeniably more powerful than the small firm, the biotechnology firm is undeniably more dependent on the large partner, and there is an imbalance of influence.

Sweeping conflicts under the rug.

- The purpose of the alliance is knowledge-creation. Such an endeavor is oblique, hard to predict, unwieldy to measure, and difficult to describe as well as to judge, and it demands giving up "ownership" of ideas.

- Collaborating involves scientists who have conceptual frameworks, vocabularies, and discipline cultures that are very different. Scientists have been trained to be, essentially, solo contributors. Multidisciplinary teamwork and cross-organizational communication are not easily achieved. (Plus, scientists have moods and quirks like the rest of humanity.)

Fortunately, the constructs of uncertainty and equivocality are as useful in the interpersonal domain as they are in the scientific and technical. We return to these constructs, this time from the vantage of the team rather than the task. Unless all members of the alliance team have worked together previously, as a group, on a similar task (so unlikely as to be impossible), there will be high levels of *team* uncertainty and *team* equivocality, as well. The interpersonal dynamics associated with these levels are predictable and, thus, can be anticipated, planned for, and intelligently managed in the post-milestone period.

Team Uncertainty

Uncertainty reflects a difference (a gap) between the information and knowledge required to accomplish some objective and that possessed by the alliance team at a given point in time. The higher the level of uncertainty, the more information is needed. *Task* uncertainty arises because of a gap in scientific and technical information. At the start of the collaboration, the gap will be widest overall and may be especially large on the pharmaceutical side (life-cycle stage asynchrony).

Team uncertainty arises because of a gap in information about team-specific behaviors, such as reporting relationships, performance evaluation systems, work norms, and so on. Earlier, we discussed the general characteristics of the large pharmaceutical partner in terms of structure and systems, but how these will affect collaboration is information that becomes available only when the collaboration is underway. At the start of the project, people in the alliance team will have very limited information about each other. Curriculum vitae and a 2-day formal "get acquainted" meeting are not sufficient to address (among others):

- What are the reporting relationships?
- What are the two company's systems for performance evaluation? Will both partners attribute a portion of the individual's performance to team activities and accomplishments?
- How will this group of people share experimental results?

Again, we cannot provide specific answers to firm-dependent questions (contractual reporting relationships, performance evaluation, structure, culture). However, we want to emphasize the importance of reducing the level of team uncertainty by addressing those issues effectively. The project charter, if it is well-designed, should provide information on formal communicating lines, reporting relationships, and some work norms. Alliance leaders must ensure that additional effort is devoted to gathering, exchanging, and sharing data and information about these issues.

Ensuring that the team creates—and lives by—good team rules can reduce the high level of team uncertainty about working together. Good team rules define how and how often people will be informed (and about what). They articulate desired norms of communication ("open," "candid," "respectful") and of listening ("active," "respectful"). Rules describe how people will relate to each other ("as peers," "leave hats outside the door") and the values that all members will seek to uphold ("collegial," "challenging"). Team rules can, of course, be different from "company rules," or policies governing performance and decision-making. The team must distinguish between how they will *adhere to* individual company policies and how they will collectively *abide by* team rules.

Good team rules that emerge from passionate discussions about collaborating will do much to reduce the high level of team uncertainty expected in the early stages of the alliance. Ideally, these rules should be completed before the end of the emergence stage. Realistically, we understand that achieving the first milestone may take up everyone's attention and energy. Prescriptively, team rules must be completed before the work moves into the growth stage of the life cycle, or progress will be hampered.

Team Equivocality

Team equivocality represents another gap and emerges from the differing perspectives that team members have with respect to their role identities (who I am, how I like to work, how I like to relate) and role identities (what

I do to further task accomplishment) within the alliance team. Some examples of team role identities include:

- *Information gatekeeper* (responsible for scanning the external environment for relevant knowledge)
- *Linchpin* (responsible for the informal interface with other functions in the organizations)
- *Cheerleader* (responsible for morale and support)
- *Affiliator* (responsible for seeing to interpersonal issues that may get in the way of work).

Over time, these personal and role identities are worked out among team members. They cannot be prescribed ahead of time (unlike collaborative behavior, which can be prescribed by team rules). Rather, they will emerge in accordance with the personalities and interests of the people. Like team uncertainty, team equivocality should be reduced, ideally, before the end of the emergence stage. Realistically, team building may remain unfinished until the post-milestone period. Again, unless it is completed before the growth stage, progress will be hampered.

It is important for leaders to keep in mind that team roles and team dynamics are complicated by the three concentric circles of cultures:

- Sector cultures (pharmaceutical and biotechnology, or "large established firm rooted in the chemical industry" and "small entrepreneurial startup rooted in academia")
- Organizational cultures (values and norms idiosyncratic to each company)
- Discipline cultures (organic chemistry, toxicology, pharmacology, bioinformatics, molecular modeling, pharmakokinetics, etc.).

We do not want to belabor the differences among team members, but we do want to ensure that leaders appreciate the complexity of alliance team dynamics and deal with it effectively. Role identities and team equivocality will be influenced by (1) the sector from which the team member comes, (2) the particular firm in which he or she works, (3) the academic discipline, and, of course, (4) the personality of the unique individual.

The "themes" or constructs of uncertainty and equivocality run through a crucial function of the biotechnology leadership: communication. Below, we describe the appropriate matching of communication and medium in the context of high uncertainty and high equivocality (task and team).

Project Communication: Message and Medium

Effective communication and effective project leadership are tightly inter-dependent. In fact, a number of scholars argue that one of the most important functions of project leadership is communication.[1] In this discussion, we focus on two underappreciated aspects of communication: media richness and media selection. If inappropriate media are used for messages, the effectiveness and speed by which work is accomplished will be at risk. We begin with "media richness" and then illustrate how the leadership must select media, based on uncertainty and equivocality.

Our discussion of media and message could have been placed earlier in this book (when we talked about the emergence stage, for instance). Clearly, selecting the right medium is important at any time, but we chose to address the topic here for two reasons. First, in our experience, leaders (perhaps intuitively) rely largely on face-to-face communication during the startup period of an alliance. As will be clear from this discussion, this is an appropriate choice, because the start of a collaboration is fraught with equivocality. Second, again in our experience, it is in the post-milestone period that leaders begin to rely on written communication (memos, reports, etc.), even when face-to-face interaction would be more appropriate. Our point is that leaders need a basis for choosing, among the many communication media that exist today, those media that will best fit the content and context of their message.

Media—From Lean to Rich. Media that facilitate the process of developing shared meaning and perspectives have been defined as *rich*.[2] Such media allow immediate feedback among individuals, use natural language to convey nuance, provide multiple cues about meaning, and permit the message to have a personal focus. Face-to-face communication "transmits" by words as well as facial expression, gestures, tone of voice, and

---

[1]Pinto, J., and Slevin, D. Critical Factors in Successful Project Implementation, *IEEE Transactions on Engineering Management*, Vol. 33, pp. 22–27, 1987. See also Posner, B., What It Takes To Be a Good Project Manager, *Project Management Journal*, Vol. 18, No. 1, pp. 51–54, 1987.

[2]Daft, R., and Lengel, R. Organizational Information Requirements, Media Richness and Structural Design, *Management Science*, Vol. 32, pp. 554–571, 1986.

body language and allows the quickest feedback of all media. Thus, this is the richest medium and the medium of choice when the objective is reducing equivocality. In descending order of "media richness" are interactive videoconferences, telephone conversations, and synchronous electronic written communication. Note that the latter, although allowing rapid feedback, does not convey nuance nor carry many cues about the individuals.

Media that do not affect shared meaning are referred to as *lean*. Lean media include one-way videos; voice and electronic messages to a group; bulletin board postings; PowerPoint presentations; graphics; tables of data, and so on. Lean media cannot transmit subtleties of meaning or emotion, nor do they allow for rapid feedback, explanation, or recognition of misunderstanding.

As should be clear from these definitions, lean media are appropriate and effective (i.e., will support and enhance the quality and speed of understanding) when the message is straightforward. Lean media are the choice when either task or team uncertainty is to be reduced. But, when issues are ambiguous, capable of more than one interpretation, and likely to be perceived differently among team members, rich media are required. Under conditions of high task or team equivocality, rich media should be used. The time and money needed to utilize them would be well spent.

The relationship between the level of equivocality and media choice is illustrated below. The four cells also indicate the problems that can arise, if the wrong medium is used:

	Lean Media	Rich Media
Low task or team equivocality	*Effective match*	*Overcomplication* (too many cues, too much data, too much "noise")
High task or team equivocality	*Oversimplification* (too few cues, impersonal, limited feedback)	*Effective match*

"Overcomplication" and "oversimplification" are not trivial communication problems. Faced with too few cues for a high equivocality issue, people may spend a disproportionate time to seek more information, or they may "fill in the blanks" themselves (often differently from what was intended). Faced with too many cues for a low equivocality issue, they may spend too much time sorting the wheat from the chaff (and may not "keep" the right understanding).

Both task and team uncertainty will be highest in the emergence and early growth stages of the life cycle. Both can be effectively and most efficiently reduced by means of lean media. However, task and team equivocality, also highest in these stages, can be reduced effectively and most efficiently only by means of rich media. Four major implications for successful alliance leadership are as follows:

- Plan for sufficient face-to-face meetings in the emergence and early growth stages, to permit task and team equivocality to be reduced successfully and rapidly.
- Reduce the level of task equivocality *before* undertaking the activities required to reduce the level of task uncertainty. If team members do not agree on the interpretation of experimental data or results, they cannot possibly be efficient in replicating them.
- Ensure the availability and adequacy of technologies supporting lean media, so that task and team uncertainty can be reduced successfully and rapidly.
- Make media choices that are consistent with the level of equivocality of the issues.

In the remainder of this chapter, we address particular leadership challenges of the growth and maturity stages of the alliance life cycle. Our emphasis is on the challenges of leadership style, "spikes" of uncertainty and equivocality, and conflict.

LEADERSHIP CHALLENGES IN THE GROWTH AND MATURITY STAGES

After the first critical milestone has been met, the alliance can be said to enter the growth stage of the life cycle. During this stage, the most rapid progress should be made in accomplishing the subtasks required to meet contractual objectives. Additional milestones will mark the ramp-up of work, but levels of task and team uncertainty, along with task and team equivocality, should be on a downward slope (although reduction will not be linear).

When task uncertainty has been reduced to a relatively low level, the alliance project will be in the mature stage of the life cycle. Ideally, the team will be functioning like a "well-oiled machine," and both partners will feel assured of alliance success (which, remember, is different from commercial success). Again, though, the mature stage of the life cycle will not be without bumps along the way. Like the emergence stage, the growth and maturity stages bring with them predictable challenges. In the following sections, we address these in three categories: leadership style, spikes of uncertainty and equivocality, and conflict. Although discussed separately, they are (of course) interdependent.

Leadership Style

We want to emphasize that what we mean by "leadership style" may be different from the vernacular terms used to describe either types of leaders or how a leader includes (or does not include) input from others. In the vernacular, we would call a particularly compelling and attractive leader "charismatic." In the vernacular, we would describe a leader who does not include input from others as "autocratic."

In our discussion, we are using "leadership style" to refer two fundamental activities of a leader: (1) *initiating structure*, and (2) *supporting interpersonal relationships*.[3] Initiating structure involves planning, scheduling, devising work procedures, evaluating results, and so on—activities directed toward the task. Supporting interpersonal relationships involves listening, communicating, resolving conflict, and so on—activities directed toward the individuals performing the task.

The corresponding leadership styles are defined as *task-focused* and *relationship-focused*. A task-focused style refers to work and the best way to accomplish it. A relationship-focused style refers to the affective or emotional dynamics of the people doing the work. Of course, the leader must be competent in each style. The important question is, Under what conditions is each style appropriate? Again, the concepts of uncertainty and equivocality can help us answer this question.

In the emergence and early growth stages of the alliance life cycle, we expect that levels of task uncertainty and task equivocality will be highest. As we noted earlier, task equivocality (multiple and competing interpretations of the experimental goals, results, data, etc.) must be reduced before

[3]Fiedler, F. A Theory of Leadership Effectiveness, New York: McGraw-Hill, 1967.

Task Focused
Leadership Style

Relationship Focused
Leadership Style

task uncertainty can be addressed. Rich communication media must be used, and leadership should facilitate bringing team members together for face-to-face discussions until shared meaning is achieved. During periods of equivocality reduction, the more effective leadership style will be relationship-focused, to facilitate the accomplishment of efforts needed to reach the milestone. Effective leaders will support interpersonal relationships, to facilitate the interactional processes necessary to deal with the high levels of equivocality. They will attend to interpersonal dynamics, to ensure that openness, respect, and candor characterize team behaviors. Put another way, effective equivocality reduction occurs, when leaders model these behaviors and facilitate appropriate interaction during discussions.

When the level of equivocality has been reduced, the level of uncertainty will still be high. Under conditions of high uncertainty, the more effective leadership style will be task-focused. Effective leaders will initiate structure. They will help to define criteria for evaluating alliance progress, they will identify weak points in the project plan, they will challenge team members and direct them to potential experts in the fields, and they will monitor and assess progress-to-goals. Put another way, effective uncertainty reduction occurs, when leaders structure the problem of information gathering and disseminating and make use of lean media to transmit uncertainty-reducing messages.

> High Equivocality . . . . Relationship-focused Leadership
> High Uncertainty . . . . Task-focused Leadership

Although leadership style should vary in response to uncertainty and equivocality, it is rare that, at any given time, the situation is either uncertain or equivocal. The very nature of research efforts, and the interpersonal and task complications of working across organizational boundaries, means that the alliance project is often facing conditions of high uncertainty and high equivocality. Effective leadership is even more challenging than our prior discussion might have implied. It is not enough for alliance leaders to be competent in each style—they will often find it necessary to be task-focused and relationship-focused in the same situation. From the perspective of uncertainty and equivocality, an effective leader must be able to recognize the changes and the differences in situations and respond appropriately.

Understanding the differences in style, and the conditions under which each is more effective, is crucial for dealing with the second category of challenges in the growth stage and maturity stages: *spikes* of uncertainty and equivocality. We also discuss why they occur, how they manifest themselves, and their implications for the alliance team and the alliance leaders.

Uncertainty and Equivocality "Spikes"

The idealized representation of the uncertainty and equivocality curves are downward sloping lines from project start to project finish. But, the reality is more complicated. First, there are two types of uncertainty and equivocality: task and team. Second, there will be spikes in each type over the course of the alliance, as we describe below. In reality, the curves (four) will probably resemble an electrocardiogram (downward sloping and without the regularity). We address each in the following discussion.

Task Uncertainty. Spikes of task uncertainty occur for two reasons. The first is endemic to science and, especially, to the fast-moving biomedical sciences. Discovery efforts are characterized by areas of imperfect knowledge, no matter how detailed the research plan or how well the research effort is going. Throughout the effort, scientists will become aware of gaps between what they know and what they need to know to move the project forward. Imperfect knowledge that impedes progress (e.g., when experimental results are unexpected or inconclusive) represents a spike of task uncertainty.

Effective leaders watch for the subtle cues that progress is faltering, instead of waiting for a formal review meeting. The advice of a biotechnology project manager, described earlier, is worth repeating.

> I have learned that I cannot stay too much on top of the work. . . . I don't need all the scientific details, because I know where morale is or not by my people's body language.

He told us that one sign of difficulty (that is, of a rise in the level of task uncertainty) was a series of empty places where the bench scientists usually worked. If people were buried in journals, sitting at their desks reviewing data for a lengthy period, or off in twos or threes looking at results, he knew he had to inquire about problems:

> Once I know the people, I understand how they telegraph their enthusiasm. In chemistry, I judge morale by the ratio of people at the hood versus at the desk. Better at the hood! In biology, people usually want to talk to me about what is going on in the experiments. If they do not want to talk to me, that is a bad sign. Another bad sign is talking too much about what won't work.

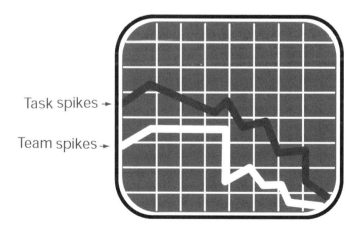

Task spikes →

Team spikes →

Technical difficulties and unexpected results (spikes of task uncertainty) are part of doing business. An effective leader does not micromanage the scientists' activities.

We had a person heading research who was described as a "femto" manager, because he managed so minutely. If there was a cell biology experiment, he had to know what was in the medium, where you bought it, how long it took the cells to grow. People like that can really put the brakes on the work (biotechnology scientist).

Rather, the leader watches for signs that the level of task uncertainty is rising. Using a task-focused style, an effective leader then structures the problem of information gathering and dissemination and makes appropriate use of lean media in the uncertainty-reduction process. The effective leader understands that "When morale rises, the research is going well again."

The second reason that task uncertainty will spike is the loss of a team member, because of transfer or resignation. Every person who leaves takes both explicit knowledge about the project and tacit knowledge. Of course, an effective leader ensures that the explicit knowledge is captured and made available to the team before the member leaves. The tacit knowledge, by definition, may or may not have been "absorbed" by people who had worked with that team member.

Metaphorically, loss of a team member leaves an individually sized gap between the knowledge the team had with the team member and the knowledge available after he or she leaves the team. In other words, task uncertainty will predictably rise with the loss of a team member. When that occurs, the project leader must oversee an assessment of progress to date, rapidly involve outside experts if needed (at least to fill the gap in explicit knowledge), and make sure that the new or replacement team member is brought up to speed as quickly as possible.

Whether the spike of uncertainty arises in the normal course of discovery or occurs because a team member leaves, the approach is the same. Using a task-focused style, the effective leader structures the problem of information gathering and disseminating and exploits the efficiencies of lean media to transmit uncertainty-reducing messages.

Both problematic scientific results and changes in team membership can cause spikes in task uncertainty. As we discuss below, they can also cause spikes in task equivocality.

Task Equivocality. It might appear improbable that the shared meaning of the alliance task, including interpretations of results, should change over the life cycle. Intuitively, if task equivocality has been reduced at the start of the project, there should be no more disagreement, because a common understanding has been reached. There are two sources of task equivocality spikes at this stage in the life cycle: results and people.

When unexpected or contrary results are obtained in the course of the planned experiments, uncertainty rises (as we have already discussed). Sometimes, equivocality will rise at the same time. This will happen when the anomalous result is so unexpected that people question not only "What should we do differently?" but also "What does this mean?" When results are open to different interpretations and when team members disagree about what the results mean for the project, then the level of task equivocality is high.

The second source of task equivocality is a change in team membership or composition. If the alliance team remained unchanged, with no new members, there would be no rise in task equivocality. But, how rarely teams stay the same over time. People change jobs, get sick, transfer to new locations. Company priorities change and more people are assigned to the project, and so on. Leaders can expect that new members from both partner organizations will join the alliance team. Each time a new individual joins the team, task equivocality will spike.

We want to emphasize that each new individual causes a rise in task equivocality. The more powerful and influential the new member, the more disruptive his or her entry into the team will be and the greater the impact on task equivocality. Every individual who is new to the alliance brings an interpretation about the scientific and technical activities and about the project plan—that is, by definition, somewhat different from the interpretation other members now hold.

Effective integration of new personnel reduces the spike of task equivocality. This is accomplished by ensuring that at least a few "old" members of the team spend face time with every new member, listening for areas of difference and conveying the team's common ground as effectively as possible. Orientation of new members is just that—their *compass* (i.e., perspective and interpretation) with respect to the project must be adjusted by communication, using rich media and facilitated by a relationship-focused leadership style. Otherwise, the explicit and tacit knowledge they should contribute may be lost.

Team Uncertainty. One of the most common reasons for spikes in the level of team uncertainty (i.e., a gap in information about team-specific behaviors) is a change in senior management:

> Another aspect of big Pharma collaborations for us is the change of people who have an impact on the team. When we ask, "where did so-and-so go?" we never receive a satisfactory answer. There are power shifts and changes in management of these large companies. We call these "undercurrents" in team interactions (biotechnology project manager).

In a small company, there is likely to be relatively little change in senior leadership over the course of an alliance. (All sink or swim with the ship.) But, in a large firm, it is not unusual for a person who was involved in crafting a deal to be promoted or to move to another part of the organization. Such movement represents the expected career path in big companies. However, these "undercurrents" cause a spike in uncertainty about formal lines of communication, partner reporting relationships, and so on, with respect to the alliance team. Reducing team uncertainty follows the same model as reducing task uncertainty. Using a task-focused style, the effective leader structures the problem of information gathering and disseminating and makes appropriate use of lean media. Even if the team is not happy with the answer to the question about where "X" went, the leader must work to see that information about "who reports to whom" is shared widely and quickly.

Team Equivocality. Every new member of the alliance team brings different perspectives about the project *and* different assumptions about individual role identity within the team. In the latter case, there are several scenarios. First, the new team member may have enjoyed the role of gatekeeper (for example) in a previous project, but may find that this role is "taken," so to speak. Second, the new team member may have no strong role preference but discover that he or she is regarded as occupying a role for which he or she feels less competent (for example). Third, the new team member may not want to be part of the team, let alone play an important role in team dynamics. Of course, there are combinations and permutations of these instances.

Once again we cannot provide specific answers, in this case to person-dependent issues. We can, however, emphasize that equivocality—about role, role competence, or taking a role—can only be reduced by commu-

The leadership challenge of conflict.

nication, using rich media, and facilitated by a relationship-focused leadership style. New team members can be wonderful resources, if they are effectively integrated by equivocality reducing efforts. They cannot simply be added to distribution lists and invited to meetings, given materials to read, told where activities are on the project timeline, and assigned to tasks.

Conflict

The third category of predictable leadership challenges over the growth and maturity stages is conflict. Studies of projects have revealed that conflict over certain issues can be expected at each stage of the life cycle.[4] Over all types of projects, the most intense conflict in the emergence and growth stages is over project priorities. The outcome of the alliance will have enormous impact on the biotechnology company, and biotechnology leadership will assign much higher priority to the alliance than will the large pharmaceutical partner. If resources are shifted away from the project, even for a short time, the consequences will be acutely felt on the biotechnology side. The resulting conflict over priorities cannot help but strain this relationship.

[4]Thamhain, H., and Wilemon, C. Conflict Management in Project Life Cycles, *Sloan Management Review*, pp. 31–50, 1975.

In the mature stage of the life cycle, the most intense conflict is over the project schedule. Not surprisingly, any delays resulting from conflict over priorities will cumulate and delay project completion. Neither cause for conflict may be avoidable, but each is predictable. Effective leadership will anticipate, plan for, and strive to manage these conflicts, so that the success of the alliance is not compromised.

We conclude with a discussion of the four leadership roles and the tasks and activities that each should accomplish during the growth and maturity stages of the alliance life cycle.

LEADERSHIP ROLES IN GROWTH AND MATURITY

Commitment Champion

The Commitment Champion faces several pressures. If this remains the only alliance in which the biotechnology company is engaged, morale throughout the company is fragile. The Commitment Champion needs to keep people focused and excited about the one alliance they have. To the extent that people in the company stay enthusiastic, they will be able to parlay their success and energy into another relationship when this one ends.

If more recent alliance deals have been signed, the Commitment Champion faces the prospect of having people's interest diverted. Bench scientists working on the "old" research may view the potential new work as more exciting. The Commitment Champion must keep the organization interested, or at least behaving as if they are.

Process Champion

The Process Champion's focus during these stages must be on maintaining effectiveness (i.e., good process) within the alliance team. This will entail working closely with the Science and Technology Champion, described below, to help deal with the spikes of task uncertainty and task equivocality. In addition, the Process Champion must be alert to issues of team uncertainty and team equivocality. The person in this role needs well-developed skills in communication and conflict resolution, to help smooth out the bumps along the way.

Science and Technology Champion

Immediately following the achievement of the first milestone, there will be a natural inclination to reduce the pace of work. When the milestone frenzy is over, team members may need—and should get—some kind of break. But, the reduction in pace cannot last too long. The Science and Technology Champion needs to focus the team on disciplined and directed research.

In the growth and maturity stages of the life cycle, scientists may lose their motivation for one of two reasons. First, the work may be going so well that it becomes routine and unexciting. Second, the work may be going poorly, with results that make no sense or results that are contradictory. Sorting out the problem might require abandoning a set of experiments or, at minimum, doing a fair amount of additional work to produce data that are usable.

When the science begins to be less motivating than before, the Science and Technology Champion must create the focus and direction the team needs to keep going. In addition, when the team experiences a spike of uncertainty, it is the Science and Technology Champion who must help structure the uncertainty reduction process. This subject matter expert must help the team decide, for example, what the critical experiments would be and what different results might mean for the next steps.

Alliance Team Members

The greater the number of team members who approach their work as leaders—of the science, of morale, of motivation—the better. The team will weather the inevitable scientific problems and setbacks, if at least some members take their leadership responsibilities seriously and are models for others to follow.

CLOSING THOUGHTS

Under conditions of high task or team uncertainty, team members may disagree on a course of action or on what the team's next steps should be. When the uncertainty is reduced by gathering and processing relevant information, the decision-making process can move forward in a rational way, and conflict should subside.

Under conditions of high task or team equivocality, however, the situation is more complex. Equivocality stems from multiple and competing interpretations of an issue. In turn, different interpretations stem from perceptions that have been filtered through team members' discipline, organizational, and sector cultures, as well as personality. People who share common training and experience will share a common way of looking at the research and of responding to scientific and technical issues. Thus, there may be conflict between disciplines and specializations, because team members will look at issues, interpret them, and communicate about them differently. The effective leader appreciates these potential complications and is prepared to address them.

Think differently . . .
Lead differently . . .
Make alliances work.

CHAPTER 10

ENDING: COMPLETION OR TERMINATION

Concluding an alliance, because the goals are met or because one partner terminates the deal, is perhaps as difficult as starting an alliance. And, it is as important to end well as to begin well—in order to do well or better the next time, whether with the same or different partner. In this chapter we focus on key issues and challenges of the final stage of the alliance life cycle (completion, the best case), and those of termination at any stage (the worst case).

"BREAKING UP IS HARD TO DO . . ."

In many ways, ending an alliance is as complicated and precarious a situation as beginning one, and the final meeting(s) is as crucial for future relationships as the first meeting. We called the first formal meeting a *critical junction* in the alliance process. The same term applies to both completion, because the goals are fully or partially met, and termination, because one partner decides to cancel the deal. If the collaboration ends well, even if the scientific and technical work is terminated, the next alliance (with the same or a different partner) is more likely to be effective than if the collaboration ends badly.

The best "best case" for concluding an alliance is the situation in which goals are met and another alliance is about to begin *with the same partner*. For the biotechnology company, this outcome means less equivocality,

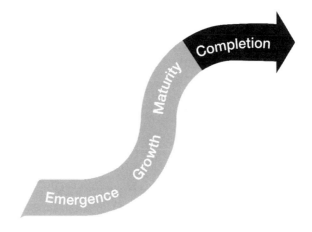

uncertainty, and anxiety at the start of the new alliance (and provides the company with much needed alliance revenue). Because the partner is the same, much of the organizational and interpersonal due diligence has already been accomplished, although scientific due diligence will be required for the new project. People on the alliance team will be confident of moving forward rapidly; they generally know each other and have worked well together. Both parties also have alliance experience. Thus, they will understand better how to define a realistic project work plan, including milestones and expenses. Scientists will be closer to a common interpretation of the research effort, because they have collaborated. Even if there are some new members of the team (very likely), task equivocality can be reduced more efficiently. Initially, of course, there will be a high level of task uncertainty associated with the new project.

The next "best case," and the one we expect most of the time, is the situation in which the alliance is completed to the satisfaction of both partners, but there is not another collaboration in the immediate future with the same partner. In this case, a successfully completed alliance is readily leveraged by the biotechnology company. Word-of-mouth will be positive, and many useful contacts will have been made in and by means of the large partner. People in the biotechnology company will be more confident in their collaborative abilities. They will also be more confident in seeking another alliance (taking the "show" on the road). Having undertaken the requisite debriefing, they have learned what they did well in the just completed alliance and what they need to improve for the next time.

At least from the perspective of the biotechnology company, the "worst case" is the situation in which the alliance is terminated by the large partner at some point between the first formal meeting and completion of contract objectives. The large partner may have decided to shift resources to another project in the portfolio. Perhaps both biotechnology and pharmaceutical scientists determined that the project was problematic, scientifically or technologically, or simply failed to achieve project goals. Sometimes, the large partner is dissuaded by scientific data and technical difficulties and is determined to bail out quickly. Whatever the reasons for termination, the impact on the biotechnology company will be sizable. Leaders will have to do "damage control," both internally and in the public eye. Company scientists have to be reassured and, at the same time, exhorted to examine what happened. Learning from termination is even more important than learning from success, albeit much more painful.

This chapter is focused on ending the collaboration. We address, below, the best case (completing the objectives) and discuss the predictable issues and challenges of the fourth stage of the alliance life cycle. We then address the termination process as a special (and, we hope, rare) instance of ending an alliance.

COMPLETION

Paradoxically, the more successful the scientific effort and the better the interpersonal interaction among team members, the harder it will be to end the collaboration. There are two critical leadership challenges of the final stage of the life cycle: (1) handing over work to the large partner, to carry the project to the marketing phase; and (2) a steep rise in team equivocality. Both challenges are expected. They must be planned for and intelligently managed. Other challenges of this stage we discuss under "leadership roles."

The Challenge of Handing Over

Successful completion of an alliance may suggest a hand-*off*, by the biotechnology side, of a final package. The big pharmaceutical partner will then, with appropriate resources, complete development and market the product.

Handing over the work.

In fact, there will be a number of times when work will be handed *over* during the mature stage of the alliance life cycle. Every late-stage milestone represents the handing over of a part of the project that is (by definition) progressively less uncertain. The "real" curve of scientific and technical uncertainty never reaches the abscissa. But, accomplishing the contract objectives implies reduction of task uncertainty to the satisfaction of both parties. The formal completion of the alliance is less an actual stage of the life cycle than a "punch list" in which both parties ensure there are no loose ends.

Two issues complicate the process of handing over the work during the conclusion of an alliance. The first has to do with the nature of science; the second is interdependent with that but has more to do with the nature of human beings.

The Nature of Science. Particularly in the rapidly moving field of biomedicine, scientific and technical achievements will prompt the scientists to expand their focus, to address new questions, and to elaborate hypotheses—in short, to continue working:

> It is always the natural inclination of scientists to refine something, to make it better. I realize, from a business perspective, that we have to stop. But, from a scientific perspective, stopping does not feel realistic to the people on the project (biotechnology CEO).

The motivation to continue working on the science is likely to be quite strong on the biotechnology side, because this environment is literally and figuratively closer to an academic environment. Depending on contract terms, biotechnology scientists may have already published in academic journals and/or presented at scientific conferences. Their colleagues in the respective "invisible colleges" will be interested in the alliance findings and the implications. Scientists will be invited to speak at future conferences and meetings.

We do not mean to imply that scientists in the pharmaceutical company will be immune to this desire to keep working. We expect similar dynamics in both firms. As a research director in a large pharmaceutical company noted:

> If you leave scientists to run their own timetable, they are like gold prospectors: "Around the corner we'll strike an even *richer* vein!"

In fact, it may make eminent sense to continue working. Preparing analogs of an active drug may permit broader patent coverage. Additional medical indications or formulations may be discovered, which also broaden potential sales of the product. The decision of when to stop, in the pharmaceutical company, should be made by senior leaders and is not dictated by the contract with the biotechnology firm.

The Nature of Human Beings. The second issue complicating the handing over derives more from the nature of human beings. The research effort will have become intensely personal. Many of our interviewees on the biotechnology side used images of *babies* and *children* when they spoke about completing the work:

> It is very hard handing over what feels like "control" to your partner. It is like handing over your children! Men and women in this company have spent their professional lives working on the project, and they worry about handing it over and not being able to touch it every day. "How is my baby doing?!" (biotechnology executive).

Scientists are likely to be frustrated and anxious about handing over the baby. Thus, people who worked well together earlier in the project life cycle may now experience problems. Those who were confident of their colleagues' expertise may now question it, as the reality of "not being able to touch [the baby] every day" becomes clearer. Trust may become an issue, simply because one "parent" is perceived to have had more experience. Candor about these topics will be hard to maintain, unless the Process Champion understands the cause and can facilitate open discussion about these feelings.

Pharmaceutical executives also appreciate the difficulty of this transition:

> We understand that if we do not manage the process well from our side, the transition of the project from mostly the biotechnology company's responsibility to mostly ours will not succeed. So, we have teams assigned to shepherd the project along, and we make sure that some of the biotechnology discovery scientists are involved as well, which is very important. That helps the transition go more smoothly.

One biotechnology project manager described the process as trying to help people "let go" of the work "while remaining joined at the hip" with the partner for the remainder of the contract. Because of both the nature of science and the nature of scientists (i.e., human beings), leadership of the handing over is more an effort of maintaining mutual trust, candor, and morale than of tying technical "knots."

Below we address the issue of morale, using the "lens" of team equivocality.

The Challenge of Team Equivocality

Throughout our discussions in this book, we have applied the constructs of uncertainty and equivocality to the alliance task and the alliance team. We noted that high levels of uncertainty and equivocality, of task and team, would be characteristic of the emergence stage. These levels would be reduced over successive stages of the life cycle, although there would be spikes.

The one exception to the general rule of "reduction over successive stages" is team equivocality. If we could plot some measure of divergence and convergence of meaning shared by the team over the life cycle, we would find a U-shaped-curve phenomenon. Team equivocality

Team Equivocality

reaches its nadir at some time in the late growth or early maturity stage and rises sharply toward the end of the mature stage of the alliance. The eruption in formerly smooth team dynamics can take unwary leaders by surprise. If not addressed effectively, the high level of team equivocality at the mature stage can jeopardize the project schedule, delay the handing over of project work, and turn colleagues into strangers. The sharp upturn in the level of team equivocality is an obstacle that can cause the alliance to founder.

Why is there a predictable rise in team equivocality during this stage of the life cycle? We believe that, in biotechnology alliances, the most compelling reason for this rise has to do with the role identities implied by the metaphor of "handing over your children." In the mature stage, a new role becomes salient: that of *parent*. Whose child is it? Who is capable of understanding the idiosyncrasies, dealing with the problems, "raising" the "child" to be marketable?

The role of parent is not limited to biotechnology team members (who, it might be added, have created the "baby"). Pharmaceutical team members struggle also:

As the invention takes shape, our scientists want to get involved, even though our contract might stipulate that the biotechnology scientists complete certain tasks. I don't think it is bad that our scientists want to take over, but we don't want to duplicate the work and raise the cost (pharmaceutical research director).

The metaphors of *child* and *parent* reflect the real and intense emotions that impact collaboration dynamics. We want to emphasize that the more successful the scientific effort, the more admirable the child. And, the better the interpersonal interaction among team members over the course of the collaboration, the more highly charged will be team emotions during completion.

Another reason for the rise in equivocality is that team members are coming to grips with the impending loss of their identity in the current alliance and are apprehensive about future collaborations. Even if there is another deal with the same company, team members may not occupy the same role in the new effort.

Ending an alliance does require addressing technical loose ends. Leaders in the biotechnology firm must reduce the relevant scientific and technical uncertainty that remains, utilizing a task-focused style that puts structure around this work. They must make use of lean media for effective and rapid communication. (Devising an electronic "punch list" is a practical illustration of both a lean medium and a means of structuring the effort.) In the pharmaceutical partner, there will be decisions about backup candidates, additional indications, better formulations, and so on. Some of the latter decisions may entail additional work with the biotechnology firm; others will be accomplished within the pharmaceutical company itself. But, the underlying question of whose ideas are more valuable (biotechnology or pharmaceutical) becomes a big issue for the alliance team and must be dealt with effectively.

At the same time, leaders must manage a high level of team equivocality that, more than the technical loose ends, can jeopardize "ending well." Intense conflict at this stage is predictably over "personality" as well as over schedules.[1] Personality-based conflict may erupt—for instance, between people who approach completing the work very differently (e.g., laissez-faire versus tightly controlling). Addressing the underlying cause of these conflicts requires a relationship-focused style, sensitivity to the overlay of parent roles on both sides of the collaboration, and intensive face-to-face discussions about what it feels like to end the collaboration. There is no shortcut to maintaining morale, trust, and candor.

---

[1]Thamhain, H., and Wilemon, C. Conflict Management in Project Life Cycles, *Sloan Management Review*, pp. 31–50, 1975.

Other Challenges

There are additional challenges, responsibilities, and tasks that face leaders in ending a completed (i.e., not terminated) alliance. Below, we organize our discussion by leadership role.

Commitment Champion. As the work is being completed, it is vital that senior executives maintain (and demonstrate) commitment until the end. The Commitment Champion will feel pulled in different directions. On the one hand, the current project must be finished. On the other hand, there may be another alliance on the horizon (or, at least, imminent "road shows" to potential partners). The Commitment Champion must work hard during this stage, to keep a sharp enough focus that the current alliance ends well. This champion must also ensure that commitment at the senior level is visible to everyone in the firm.

What is required to keep the commitment lively and visible at this stage? If resources are needed, the Commitment Champion must work to secure them. More importantly, the Commitment Champion must continue to follow the progress of the alliance personally and make sure that everyone in the company is kept apprised.

Process Champion. Without explicit intervention, this stage of an alliance will be characterized by problems of morale and motivation. It is the responsibility of the Process Champion to ensure positive team dynamics, which is not an easy task. Regardless of the scientific and technical accomplishments, the completion of work will have a large emotional impact on the project scientists, in terms of both future work and future teams. The Process Champion needs to be very sensitive to these dynamics. If morale and motivation are allowed to become problems, they can jeopardize the effectiveness (i.e., schedule) of the alliance.

Science and Technology Champion. The responsibility of this champion at the final stage is, very simply put, "all in the details." The science needs to be carefully finished and well-documented. Future due diligence efforts and patent questions require that what is done be clear, complete, and recorded in detail in laboratory notebooks.

Alliance Team Members. The responsibility of Alliance Team Members is to finish up in a timely and collegial fashion. They must complete their effort with great discipline, accuracy, and attention to detail, along with clear records of the experimental methods and results.

TERMINATION

As our case study in Chapter 1 demonstrated, a failed alliance can be devastating to the biotechnology partner:

> The chief executive of Pharma Sciences said they decided to shift their R&D allocation to compounds with greater likelihood of near-term success. Lucida researchers had described the early in vivo work as promising, but both groups subsequently found ambiguous data from Phase I studies. Pharma Sciences was concerned that the ambiguity could delay even the design of Phase II trials by 12 to 18 months. . . .

> . . . As a result, Lucida's share price dropped precipitously by more than one-third. The value of the original contract for Lucida was as much as $100 million over 3 years. Now, Lucida is no longer eligible for most of the $80 million in milestone payments.

We will assume in this section that our advice has been taken. Biotechnology leaders have conducted due diligence efforts; the first formal meeting was a success by the criteria we have listed; the team has been formed and is working well; and the technical goals are being achieved. Even in these best of circumstances, however, senior executives in a large company can change their mind about an alliance. It is simply one small item in their portfolio, and priorities may shift over time.

The primary responsibility of biotechnology leadership in this situation is *damage control*. Presenting a calm and positive public face is imperative, and it is required by everyone in the firm. The investment community will make assumptions about the value of the science and the value of the company. There must be rapid external communication by executives about the termination, in numerous press releases, press conferences, newspaper articles, telephone calls to investment contacts, and so on. These should be honest in addressing that the alliance was stopped prematurely and at the same time positive about the caliber of the science involved. Other research that the firm is conducting should be described. The objective is to ensure an impression of "strategic dispassion" (i.e., priorities have shifted in the partner firm) and intellectual competence (in the biotechnology firm). The basis for investing in the biotechnology company in the first place has not disappeared because the alliance was terminated, unless the whole company was devoted to the project that failed. (The latter instance is beyond the scope of this book.)

Leaders are also responsible for damage control internally. They must communicate immediately to everyone in the company, and it must be face-to-face. Termination is an extremely equivocal subject and requires the richest medium. We do not assume that leaders can answer "what does this mean?" to everyone's satisfaction. But, they must engage in the process of addressing what it might mean now and in the future. The meeting to discuss the termination will be very difficult for everyone. Leaders must be scrupulous in not casting blame on the partner, because "bad-mouthing" will be

Damage control when an alliance has failed.

repeated outside the company and is likely to get back to the partner. Any indication that people are pointing fingers must be confronted and stopped. The purpose of the meeting is to surface the expected feelings of anxiety and concern, to assure people that their scientific competence is valued and valuable, and to acknowledge that everyone is affected by the abrupt ending. The purpose of the meeting is not to find blame or fault or to analyze the collaboration (the alliance debrief, discussed later in this chapter, is separate and different from this meeting). Leaders must also be honest about future job security, theirs as well as the scientists'. They must walk a fine line between total pessimism in this regard and unfounded optimism.

Once senior executives have communicated face-to-face, they should use lean media for updates and additional news. People have to be kept informed, even of the negative publicity. Additional meetings may have to be called, if the turmoil (i.e., equivocality) remains so high that scientists have trouble working. This happens. People may be at the bench, but if anxiety is too intense, they cannot work productively.

Although the biotechnology side will bear the burden of termination, executives in large companies realize that their reputation as partner can suffer if the ending goes badly:

> We have to manage separation well. We try to be sensitive to how much it means to the small biotechnology company when we have to terminate an alliance. Not every alliance works out. When it does not, we have to make a clean separation but not "burn bridges."

> Maybe the next idea from this company will work. This has to be managed very carefully, or it can create a lot of problems. We are mindful that other biotechnology companies are very aware of how we treat the firm from whom we want a divorce (pharmaceutical CEO).

Other biotechnology companies will be bringing their ideas to the pharmaceutical firm. The reputation of the large firm, as a partner in alliances, will have an impact on future collaborations.

LEARNING

Learning is a vital part of effective collaboration, and a debriefing process must be part of every alliance conclusion. We are confident that everyone would agree that it is important to learn why an alliance was successful (that is, why the collaboration was effective) or why it was a failure (that is, the collaboration was terminated abruptly by the partner). Our observation and experience, however, suggest that the *doing* is harder than the *saying*, particularly when it comes to looking back in time.

If an effort goes well, it is human nature to assume that "*we* knew what we were doing." If it goes poorly, it is also human nature to look at external causes of problems ("*they* had no patience"). In other words, we are apt to attribute success to ourselves and failure to an outside influence. In addition, everyone must move on to new challenges and may feel "too busy" to analyze the just completed project.

Learning from Success and Failure

People need to know what they did that might have contributed to alliance effectiveness. Of course, all people can really do is to come up with hypotheses or propositions linking behavior and outcome. For instance, proposed reasons for alliance effectiveness include the following:

- We discussed, internally, the given asymmetry between biotechnology and pharmaceutical companies well enough to appreciate and manage the implications for our relationship.
- We undertook the right due diligence on the organization and the people; as a result, we had a good sense of our partner's strategy and the perspectives of their scientists.
- We prepared every person involved with this alliance, paying special attention to the interpersonal skills of each team member.
- We stayed in touch with the team so that we were able to monitor productivity and morale. Communication was effective and efficient at all levels. There were no surprises.

If the collaboration went badly, whether or not the alliance was terminated, then reasons for alliance ineffectiveness must be examined:

- We may not have considered the impact of the change in senior management in the large partner on the priority of our alliance.
- We may not have been sufficiently up-front with our partner about some of the difficulties we had with a set of experiments. They may have been unprepared for the time it took us to "get it right."

- We may not have trusted our partner enough to hand over our data efficiently.

Note that we have couched all propositions in the "first person," because of the lopsided leadership and management responsibility discussed throughout this book. The analysis of relationship effectiveness and ineffectiveness must be examined from the perspective, first, of the biotechnology firm. Only when all propositions about the biotechnology leadership and team have been exhausted can people turn their attention to finding cause (for positive and negative outcomes) in the partner.

The type of examination that must be conducted on an alliance follows the general framework of scientific inquiry. Hypotheses should be constructed and tested. Ideally, some measures for behaviors should be agreed upon and used in the testing of hypotheses. Consider the hypothesis: "We may not have trusted out partner enough to hand over data efficiently." Possible metrics for "insufficient trust" might include the following:

- Decrease in number of e-mails from the biotechnology scientists to their colleagues
- Spotty attendance of biotechnology team members at late-stage milestone meetings
- Notable decrease in the frequency of face-to-face discussions over time.

Ironically, many scientists find it difficult to translate the skills of the bench from science to relationships. They are trained to work individually, using their personal talent and education to construct and test theory (a closed-loop feedback and control process). However, although they may construct normative theories of behavior, our experience is that they may not seek feedback on their assumptions. The purpose of the debriefing process, the learning from success and failure, is for everyone involved in the alliances to make their assumptions explicit, to agree on measures for the appropriate parameters, and to examine the evidence of effectiveness and ineffectiveness on the biotechnology side.

The objective of the analytic exercise following every alliance is twofold: to generate explicit knowledge about collaboration (e.g., general rules that can be codified and kept in an accessible file) and to generate tacit knowledge about collaborating. A "collaborative intelligence" database should be part of the knowledge management activities in the biotechnology company, no matter how few the employees. More importantly, skilled individual collaborators are an invaluable asset to the firm. Effective collaboration requires intellectual competence as well as relational competence.

Some of Our Findings

A number of biotechnology firms with whom we have worked have conducted analyses of completed alliances, although typically they have not followed as formal a process as we would advise. We have found common themes from their efforts to learn from success and failure.

With Respect to Alliances That Were Effective and Successful

- Team members worked hard at managing the relationship. They accepted the lopsidedness and moved on. They did not spend time wishing it were otherwise. They did not spend time complaining about how unfairly responsibility was divided.
- Team members were disciplined and careful in their science. Experiments were well planned, and data were recorded accurately and completely.
- Team members did not "guard" information. They came forward quickly when there were problems, believing that early warning was better than late surprises.
- Team members were willing to work collegially with each other and with scientists from the partner organization. Their energy was focused on the project, not their own egos.

With Respect to Alliances That Were Less Than Effective and Successful

- Team members could remember a meeting or a point in time when things went sour. But, rather than work deliberately to turn things around, they assumed the problems would disappear.
- Team members soon identified a "difficult scientist" who was not managed well. As a result, the person jeopardized relationships within the team and with the partner's scientists.
- Team members felt that, most of the time, they had to fight for attention and commitment from the senior leaders. They believed that

executives did not give either their time or the resources that might have helped to maintain the alliance on the right track.

These are only highlights of what people believed the team did well or the team did poorly. In the actual debriefing process, the analysis should address not only what the team did right (assuming a successful alliance) but also what the team could have done better. If the alliance was unsuccessful, the debrief should identify what the team did well. A narrow focus only on what was done well or what was done poorly produces less-than-optimum learning.

CLOSING THOUGHTS

Collaboration must be examined with the same care as scientific data from the laboratory. Essentially, every collaboration is an experiment in the behavioral as opposed to "hard" sciences. When it is over, it is a phenomenon to be investigated as rigorously as possible. During the debriefing session, each of the four leadership roles (Commitment Champion, Process Champion, Science and Technology Champion, Alliance Team Member) is responsible for reporting what worked and what did not. To do this well, people in these roles will have been deliberately attentive and focused throughout the course of the alliance. They will have kept detailed notes and, by sharing them with the group, the entire company will benefit.

During this session, the group will need to "digest" what has been said. At times, the interpretation will be relatively straightforward. At other times, different members of the debrief group will "hear" different things in the reports that are presented. Such equivocality must be dealt with. Unless meaning and interpretation are common, the team will not learn effectively and will not be able to act appropriately on what has been learned. Once the equivocality has been dealt with, there may still be some uncertainty about the collaboration. There may be information and data that the group would like to gather or have at its disposal, to better understand what happened.

Process is very important in this as in all meetings. Each "voice" and everyone's perspectives must be heard. Common interpretations and meaning must be reached. Because the Process Champion has been an integral player in the alliance itself, most groups would do well to have an outside consultant to assist at the debriefing. There are likely to be emotionally charged issues that will arise, and dealing effectively with them may be difficult without professional help.

Think differently . . .
　　　　Lead differently . . .
　　　　　　　Make alliances work.

CHAPTER 11

READINESS, LEARNING, AND ALLIANCE EFFECTIVENESS: A ROAD MAP

What should you know, and what should you do, to promote alliance effectiveness?

Our book has exposed you to a case study of a real alliance; context; history; the experiences of many alliance participants; and such concepts as uncertainty, equivocality, tacit knowledge, the life cycle, and so forth. We expect that you will "see" alliances differently, through new frames for viewing these collaborations. We expect you to understand some of the different ways that alliance leadership is accomplished, by means of the four leadership roles.

This chapter is very simply structured. It contains a few questions, in the order of the prior chapters (hence, a road map). Moving from thinking differently (perceiving, "seeing") to leading differently and effectively should be facilitated by your addressing, individually and collectively, the questions we pose below. They constitute a tool you can use over the course of an alliance, from deal signing to completion, as a literal roadmap for effectiveness. You can also use the tool as needed, to monitor performance in an ongoing partnership and to ensure that critical issues are addressed in time.

Our questions reflect the fact that leading alliances right, right from the start, requires both *knowing and doing* and *issues and actions*. Sometimes, simply identifying the issues and reflecting on their implications prepare you to lead better. Other times, knowing is not enough. You have to act on what you know. We recommend that key people in the biotechnology firm first tackle the questions individually. It is important to determine if answers are widely divergent in any area, and to ascertain why. Moreover, responsibility for the knowledge and actions implied by the questions will not rest with a single individual. Making alliances work entails the efforts of several (if not many) people.

We also recommend that the road map be used as an opportunity to model collaboration within your own alliance leadership team. As a group, you can discuss the questions and share your answers. Ultimately, there should be general agreement on most of the major points. Interpersonal issues in your own firm that could get in the way of alliance effectiveness may emerge in this process. If so, they can be dealt with ahead of time and in house. Better to address problematic dynamics "at home" than air them in the public arena of the alliance.

Chapter 1 (The Lucida–Pharma Alliance)

1. What aspects of this case "hit close to home"? Were you surprised at the demise of Lucida?

2. If you were asked to advise Mark Santoro, when he started the job of vice president of research at Lucida, what steps would you have recommended he take?

3. If he followed your advice, what do you believe would be the major difficulties facing him?

4. How should he address those difficulties?

5. How would you have handled Will O'Brien?

Chapter 2 (Many Alliances, Many Problems)

1. Given that at least half of all biotechnology alliances experience difficulties, what are you doing (will you do) to ensure that your current (future) alliance will be effective and successful?

2. From your own first-hand experience and from the stories you have heard, do the data in this chapter under- or overestimate the problems in alliances?

3. In your own experience, what have you found to be the specific challenges to making alliances work?

4. How have you addressed those challenges? How might you addressed those same challenges in the future?

Chapter 3 (Contrasting Cultures)

1. What examples and instances can you give that illustrate how sector culture is reflected in your own firm?
2. What examples and instances can you give that illustrate how sector culture is reflected in your pharmaceutical partner?
3. What contrasts in sector cultures do you think are most difficult to deal with? Why?
4. Which contrasts or differences have had the greatest impact on alliance relationships? Why?
5. How do you think you should deal with sector culture differences that get in the way of effective relationships?

Chapter 4 (Differences and Disparities)

1. What are the most salient disparities between you and your partner?
2. How do these play out in the collaboration?
3. What are some phrases you have heard—or used—in your own company that reflect the "elephant and mouse" realities of biotechnology alliances?
4. As the "mouse," how do you feel about the assertion that it is up to the biotechnology side to make the alliance work?
5. What challenges does this leadership burden pose to you (and your company) and to the people in your pharmaceutical partner?

Chapter 5 (Preparing the Organization)

1. How would you assess your company's alliance readiness with respect to the fundamental issues of fit with your own strategy/vision and your firm's leadership commitment?
2. What has already been accomplished in the company, to prepare for collaboration?
3. What do you think still needs to be done (structure, systems)?
4. How will the critical leadership roles be accomplished for this alliance (people, combinations of roles, and so on)?

Chapter 6 (Individual and Organizational Due Diligence)

1. Who are the key alliance people in the pharmaceutical partner, regardless of formal title?
2. What do you know about them? What do you want or need to know? How will you gather the information?
3. What do you understand your partner's strategy to be? How does the alliance fit in their portfolio, at this time? How might that fit change? How will you know?
4. What evidence do you have of your partner's commitment to the outcome of the alliance? To being a good partner over time?
5. What do you know about the structure and systems of your partner? How will their structure and their systems affect alliance dynamics? How will you deal with their structure and systems?

Chapter 7 (First Meetings)

1. What informal meetings have you had, or do you plan to have, for people on both sides to meet and get to know each other and get to know the science?
2. What have you learned from the meetings that have been held that you might not have learned otherwise?
3. What are the plans for the first, formal meeting and alliance kick-off event (who, what, where, when)?
4. What should be the agenda for that meeting?
5. What are the desired outcomes of that meeting? What can you do to facilitate those outcomes?

Chapter 8 (To the First Milestone)

1. In what ways do you think the life-cycle model is useful for understanding alliance issues over time?
2. What are the implications of the potential scientific and technological asynchrony in experience between you and your partner?
3. What plans are in place to reduce this asynchrony smoothly? How can you facilitate that process?
4. What would be the evidence that alliance team members shared a common interpretation of the science and technology at this time (results, data, methodology, etc.)?

5. What scientific and technologic questions that exist at the start of the alliance could be answered by more information (i.e., uncertainty)? What scientific and technologic questions would have to be answered by face-to-face interaction and discussion (i.e., equivocality)?

6. Whose tacit knowledge is important to the success of the project? What plans are in place to facilitate the sharing of that knowledge with other team members (on both "sides")?

Chapter 9 (Growth and Maturity)

1. How do people feel after accomplishing the first milestone?

2. What issues and tensions do you believe may have been swept under the rug? How will these be addressed?

3. Do you think that appropriate media choices are made and leadership styles used, under changing conditions of uncertainty and equivocality?

4. Do you see spikes of task uncertainty or task equivocality associated with inconclusive or unexpected experimental results? Do you see such spikes associated with a new member of the alliance team? What is being done to manage these spikes?

5. How do new members, or the loss of a member, affect team uncertainty and team equivocality? What is being done to manage these changes in personnel?

Chapter 10 (Ending)

Assuming the alliance is successful . . .

1. What are the science and technology challenges in handing over work to the pharmaceutical partner? How will they be addressed?

2. What are the human issues in letting go of the work? How will they be addressed?

3. What have you learned that will make you a better partner in the next alliance?

4. What could you have done differently in this alliance?

5. How will you ensure that "lessons" are both learned and institutionalized?

If the alliance is terminated . . .

1. What do you think went wrong? Why?
2. What might be done differently next time? How will you ensure that happens?
3. What do you think went well, despite the termination?

Think differently . . .

Lead differently . . .

Make alliances work.

CHAPTER 12

IF WE COULD TURN BACK THE CLOCK . . .
(A HYPOTHETICAL CODA TO THE LUCIDA–PHARMA SCIENCES CASE)

October 1996

It is the fall of 1996, and Phil Dean (CEO of Pharma Sciences) has just left a meeting of the Board of Directors of Lucida Biotech. Back in his office at Pharma Sciences, he calls his COO and asks him to drop by. H. Ross Johnson sits down, and the two of them discuss the potential of a collaboration with Lucida. Both people know Lucida's CEO, Geoff Pitchly. Johnson is friends with Geoff; Dean is on the company board.

Dean and Johnson like the idea of an alliance. Dean charges Johnson with ensuring that the VP of Research at Pharma Sciences takes the necessary preparatory steps (identifying the likely alliance champion and overseeing the three key tasks). Both senior executives value their firm's reputation as a good partner and want to keep that reputation.

A few weeks later, Dean telephones Pitchly at Lucida, to set up a more formal meeting about a possible alliance. . . .

Late Winter 1996

Pitchly realizes he has a lot of work to complete, before Lucida is ready to partner with Pharma Sciences. First, he must hire a VP of Research, who will be put in charge of all alliances and in charge of Will O'Brien. Second,

he has to assess the likely strategic fit of Lucida's technology in the portfolio of Pharma Sciences.

After starting the hiring process for VP, Pitchly calls his senior management team to a series of meetings, to discuss the troubling data on biotechnology alliances in general (Chapter 2). In the meeting, he and his team identify just how different Pharma Sciences is from Lucida and what the implications might be for collaboration (Chapters 3 and 4). The VP of Strategy volunteers to find out more about Pharma Sciences' culture as well. Dean has the reputation for tough-minded management, and Lucida executives want to be prepared for what that might mean for them. . . .

Early Spring 1997

The intense search effort for a VP of Research at Lucida pays off: Mark Santoro comes on board, just as more formal negotiations are underway between Lucida and Pharma Sciences. Santoro soon hires a scientist with relevant subject matter expertise and project management training and experience. This person he charges with overseeing the requisite due diligence process (Chapter 6). Santoro also realizes that Will O'Brien might be difficult to engage in the collaborative processes required by the upcoming alliance (Chapter 8). Thus, Santoro begins exploring a number of options, including treating O'Brien as a valued but virtual member of any alliance team.

Lucida's new project manager gets in touch with her counterpart at Pharma Sciences. They discuss some of the scientific and technical issues involved in what will be the focus of the alliance. A National Institutes of Health symposium on the science is scheduled for late March, and both project managers decide that they and a core group from each firm will attend. This symposium, they believe, will better prepare the alliance team for the technical complexity they face in this area. . . .

June 1997

Local and national papers carry the story of the Lucida–Pharma Sciences deal. Lucida's stock price rises at the news of the milestone payments that this alliance can provide. Pitchly and Santoro, however, realize that the hard part is just beginning. They know they have to bear the larger burden of responsibility for alliance effectiveness and success. All the key people at Lucida attend a 2-day offsite meeting, to discuss and plan how they will think differently, lead differently, and make *this* alliance work. . . .

Think differently . . .

        Lead differently . . .

                Make alliances work.

INDEX